The Biological Mind

The Biological Mind

HOW BRAIN, BODY, AND ENVIRONMENT

COLLABORATE TO MAKE US WHO WE ARE

ALAN JASANOFF

BASIC BOOKS

NEW YORK

Basic Books
Hachette Book Group
1290 Avenue of the Americas, New York, NY 10104
www.basicbooks.com

Printed in the United States of America

First Edition: March 2018
Published by Basic Books, an imprint of Perseus Books, LLC, a subsidiary of Hachette Book Group, Inc. The Basic Books name and logo is a trademark of the Hachette Book Group.

Print book interior design by Linda Mark.

Library of Congress Cataloging-in-Publication Data
Names: Jasanoff, Alan, author.
Title: The biological mind : how brain, body, and environment collaborate to make us who we are / Alan Jasanoff.
Description: New York : Basic Books, 2018. | Includes bibliographical references and index.
Identifiers: LCCN 2017052705 (print) | LCCN 2017055145 (ebook) | ISBN 9781541644311 (ebook) | ISBN 9780465052684 (hardback)
Subjects: LCSH: Neurosciences. | Psychophysiology. | BISAC: SCIENCE / Life Sciences / Neuroscience. | PSYCHOLOGY / Cognitive Psychology. | MEDICAL / Neuroscience.
Classification: LCC RC321 (ebook) | LCC RC321 .J37 2018 (print) | DDC 612.8/233—dc23
LC record available at https://lccn.loc.gov/2017052705
ISBNs: 978-0-465-05268-4 (hardcover); 978-1-5416-4431-1 (ebook)

LSC-C

10 9 8 7 6 5 4 3 2 1

to Luba and Nina,
who make me who I am

CONTENTS

INTRODUCTION

Wᴴᴬᵀ ᴹᴬᴷᴱˢ ʏᴏᴜ *YOU*?
Wherever you come from and whatever you believe about yourself, chances are that to some extent you know your brain is the heart of the matter. Although it is said that there are no atheists in foxholes, there are also few people who will not duck when the shooting starts—nobody wants a bullet in their brain. If you trip and fall forward on a concrete sidewalk, your arms rise instinctively to protect your head. If you are a cyclist, the only protective gear you probably wear is your helmet. You know something important is under there, and you will do what it takes to keep it safe.

Your concern for your brain probably does not end there. If you are smart or successful, you pride yourself on your brainpower. If you are an athlete, you prize your coordination and stamina, likewise products (at least in part) of your brain. If you are a parent, you worry about your child's brain health, development, and training. If you are a grandparent, you may worry about your own aging brain and the consequences of brain atrophy. If you had to swap body parts with someone else, your

1

brain would probably be the last part you would consider exchanging. You identify with your brain.

How complete should this identification be? Is it possible that everything truly significant about you is in your brain—that in effect, you *are* your brain? A famous philosophical thought experiment asks you to consider just this possibility. In the experiment, you imagine that an evil genius has secretly removed the brain from your body and placed it in a vat of chemicals that keeps it alive. The brain's loose ends are connected to a computer that simulates your experiences as if everything were normal. Although this scenario seems like nothing more than science fiction, serious scholars use it to consider the possibility that the things you perceive may not in fact represent an objective reality outside your brain. Regardless of the outcome, the premise of the thought experiment itself is that being a brain in a vat violates no physical principles and that it is at least theoretically conceivable. If scientific advances eventually made it possible to maintain your disembodied brain, the scenario implies that the irreducible *you* would indeed be in there.

For some, the idea that people can be reduced to their brains sounds a powerful call to action. A young woman named Kim Suozzi heard that call. At just twenty-three years old, Suozzi was dying of cancer, but she refused to go gentle into that good night. She and her boyfriend decided to raise $80,000 in order to fund the preservation of her brain after she died. Suozzi believed that technology might one day enable her to be brought back to life, either physically or digitally, through structural analysis of her frozen organ. Science is nowhere near up to the task right now, but that did not deter her. To Suozzi in her final days, the brain became everything. Others have taken Suozzi's path as well. I myself have had a related experience, which I will describe later in this book.

When we are confronted with mounting evidence that the brain is central to all we once associated with our selves, our spirits, and our souls, it is not surprising that some of us react dramatically. In our brave new neuroscientifically informed world, the brain bears the legacy of several millennia of existential angst. Our ultimate hopes and fears can come to revolve around this organ, and in it we may seek answers to eternal questions about life and death, virtue and sin, justice and punishment. There is no mental function for which researchers have not succeeded in finding corresponding activity patterns in the brain, using either imaging techniques in people or more invasive measurements in animals. We see brain data increasingly entering courtrooms, the risk of brain injury newly affecting our pastimes, and brain-targeted medicines prescribed to alter a gamut of behavior from school performance to social graces. A lesson from the legendary Greek philosopher Hippocrates is penetrating the public consciousness: "Men ought to know that from nothing else but the brain come joys, delights, laughter and sports, and sorrows, griefs, despondency, and lamentations."

Everything important about us seems to boil down to our brains. This is a stark claim, and my aim in this book is to show that it sends us in the wrong direction, by masking the true nature of our biological minds. I argue that the perception that the brain is all that matters arises from a false idealization of this organ and its singular significance—a phenomenon I call the *cerebral mystique*. This mystique protects age-old conceptions about the differences between mind and body, free will, and the nature of human individuality. It is expressed in multiple forms, ranging from ubiquitous depictions of supernatural, ultrasophisticated brains in fiction and media to more sober scientifically supported conceptions of cognitive function that emphasize inorganic qualities or confine mental processes within neural structures. Idealization of the brain infects laypeople and scientists alike (including myself), and it is compatible with both spiritual and materialist worldviews.

A positive consequence of the cerebral mystique is that exalting the brain can help drive public interest in neurobiological research, a tremendous and worthy goal. On the other hand, the apotheosis of the brain ironically obscures consequences of the most fundamental discovery of neuroscience: that our minds are biologically based, rooted in banal physiological processes, and subject to all the laws of nature. By mythologizing the brain, we divorce it from the body and the environment, and we lose sight of the interdependent nature of our world. These are the problems I want to address.

In the first part of this book, I will describe the cerebral mystique as it exists today. I will do this by considering themes in today's neuroscience and its public interpretation that underemphasize the brain's organic, integrated characteristics. I argue that these themes promote a *brain-body distinction* that recapitulates the well-known mind-body dualism that dominated Western philosophy and religion for hundreds of years. By perceiving virtual barriers between our brains and our bodies—and by extension between our brains and the rest of the world—we see people as more independent and self-motivated than they truly are, and we minimize the connections that bind us to each other and to the environment around us. The disconnected brain acts as a stand-in for the ethereal soul, inspiring people like Kim Suozzi to preserve their brains upon death in the hope of attaining a form of immortality. In upholding the brain-body distinction, the cerebral mystique also contributes to chauvinistic attitudes about our brains, minds, and selves, such as the egotism of successful leaders and professionals and the "us versus them" attitudes of war and politics.

In the individual chapters of Part 1, I will introduce five specific themes that give rise to the brain-body distinction and that tend to elevate the brain above the rest of the natural realm. By scrolling through alternative, scientifically grounded perspectives, I will try to bring the brain back down to earth. The first theme I will address is *abstraction*, a tendency for people to view the brain as an abiotic machine based on fundamentally different principles from other living entities. This is best exemplified by the familiar analogy of the brain to a computer, a solid-state device that can be perfected and propagated in ways that evoke a disembodied spirit. The second theme is *complexification*, a vision of the brain as so vastly complicated as to defy analysis or understanding. The inscrutably complex brain is a convenient hiding place for mental capabilities we want to possess but cannot explain, like free will. The third theme is *compartmentalization*, a view that stresses the localization of cognitive functions without offering deeper explanations. Supported largely by the kinds of brain imaging studies we often see in the media, the compartmentalized view often facilitates shallow interpretations of how the brain helps us think and act. The fourth theme is *bodily isolation*, a tendency to see the brain as piloting the body on its own, with minimal influence from biological processes outside the skull. The fifth and final theme is *autonomy*, the view of the brain as self-governing, receptive to the environment but always in control. These last two themes allow us to see ourselves as cut off from impersonal driving forces both inside and outside our bodies that nevertheless dramatically affect our behavior.

In Part 2, I will explain why a more biologically realistic view of our brains and minds is important, and how it could improve our world. I consider three areas that today are heavily influenced by the cerebral mystique: psychology, medicine, and technology. In psychology, the mystique fosters a view that the brain is the prime mover of our thoughts and actions. As we seek to understand human conduct, we often think first of brain-related causes and pay less attention to factors outside the head. This leads us to overemphasize the role of individuals and underemphasize the role of contexts in a range of cultural phenomena, from criminal justice to creative innovation. An updated view that moves beyond idealizations must accept that the body's physiological milieu, encompassing but not bounded by the brain, provides an unequivocal meeting point for influences both internal and external to every person. Our brains seen in this way are complex relay points for innumerable inputs, rather than command centers endowed with true self-determination. Whenever I have an idea, my idea is the product of all of these inputs converging at once around my head, rather than mine alone. When I steal or kill, whatever happens in my criminal brain is the product of my physiology and environment, my history, and my society, including you.

In medicine, a grave consequence of the cerebral mystique is to perpetuate the stigma of psychiatric disease. Accepting that our minds have a physical basis relieves us of the traditional tendency to view mental illnesses as moral failings, but recasting psychiatric conditions as brain disorders can be almost as damning to the patients affected. Society tends to view "broken brains" as less curable than moral flaws, and people thought to have problems with their brains can be subject to greater suspicion as a result. Equating mental disorders with brain dysfunction also skews the treatments people seek, leading to greater reliance on medications and less interest in behavioral interventions such as talk therapy. And seeing mental illnesses purely as brain diseases overlooks an even deeper issue—the fact that mental pathologies themselves are often subjectively defined and culturally relative. We cannot properly grapple with these complexities if we reduce problems of the mind to problems of the brain alone.

For some people, the cerebral mystique inspires technological visions for the future. Many of these revolve around science fiction and the idea of "hacking the brain" to improve intelligence or even eventually upload our minds and preserve them for eternity. But the reality of brain hacking is less glamorous than its image. Invasive brain procedures have historically incurred high risk of injury and helped only the most debilitated patients. The neurotechnological innovations that meet society's needs might best remain outside our heads; indeed, such peripheral tech is already turning us into transhumans armed with portable and wearable electronics. Both hopes and fears about neurotechnology are distorted by artificial distinctions between improvements that work directly and those that work indirectly on our central nervous systems. By demystifying the brain we will be better able to enhance our lives while solving the scientific and ethical challenges that arise along the way.

Before getting into my argument, I want to say a few words about what this book does *not* try to do. First, it does not explain how the brain works. Unlike many other authors, I am concerned more with what the brain *is* than what it does. Although several of my chapters include examples of specific brain mechanisms, my purpose in introducing them is largely to illustrate modes of action that depart from widespread stereotypes about the brain. Just as many artists strive to give emotional and psychological depth to flat figures from history and legend, I hope in a humble way to add dimensionality and nuance to an organ that popular writing often depicts as a dry computing machine rather than a thing of flesh and blood.

Second, this book does not challenge the fact that the brain is essential to human behavior. Functions of the mind all require the brain, even if they do not reduce to the brain. Many of these functions are almost as poorly understood now as they were fifty or a hundred years ago, and basic neuroscientific explorations of phenomena such as memory, perception, language, and consciousness are the best way to advance our knowledge. I will illustrate how traditional ways of looking at the brain can be complemented by alternative and broadened views, but neuroscience and the brain remain at the center of the picture.

Third and most important, this book in no way aims to reject objective neurobiological findings. The perspectives I offer will foster a view of our minds and selves as more interconnected than Old Age culture traditionally views them, but this is no invitation to slip into ungrounded New Age spirituality. It is hard scientific research itself that paints a picture of the brain as biologically grounded and integrated into our bodies and environments. Conversely, it is the cerebral mystique and its emphasis on the extraordinary features of brains that drive people to doubt the power of science to illuminate human thought and behavior—a view that I, like most neuroscientists, emphatically reject. The cerebral mystique limits the impact of neuroscience in society today by presenting the brain as a self-contained embodiment of the mind or soul. This view makes it easier to "black-box" the nervous system, to treat what happens in the brain as confined to the brain, and to ignore what neuroscience might have to say about real-world problems. This is a view I mean to set aside, and I hope that this book will convince you to agree.

part I

THE CEREBRAL MYSTIQUE

one

EATING THE BRAIN

W HEN I FIRST TOUCHED A BRAIN, IT WAS BRAISED AND enveloped in a blanket of beaten eggs. That brain had started its life in the head of a calf, but ended in my mouth, accompanied by some potatoes and a beverage at an economical eatery in Seville. Seville is a Spanish city famous for its tapas, and *tortilla de sesos*, as well as other brain preparations, are occasional offerings. On my brain-eating trip to Seville, I was too poor to afford sophisticated gastronomic experiences. Indeed, some of my most vivid recollections of the trip included scrounging around supermarkets for rather less satisfying food, while the delectable tapas remained out of reach, only for the ogling. The brain omelet was certainly one of the better meals I had.

My next encounter with *sesos* came many years later in a laboratory at MIT, in a crash course on neuroanatomy whose highlight was certainly the handling and dissection of a real sheep's brain. At that time, I was drawn to the class and to the sheep's brain by a diffuse set of concerns that motivate many of my fellow humans to follow and even embed themselves in neuroscience. The brain is the seat of the soul, the mechanism of the mind, I thought; by studying it, we can learn the secrets of cognition, perception, and motivation. Above all, we can gain an understanding of ourselves.

The experience of handling a brain can be awesome, in the classical sense of the word. Is this lump of putty really the control center of a highly developed organism? Is this where the magic happens? Animals have had brains or brain-like structures for nearly five hundred million years; over 80 percent of that time, the ancestors of sheep were also our ancestors, and their brains were one and the same. Reflecting that extensive shared heritage, the shape, color, and texture of the sheep's brain are quite like our own, and it is not hard to imagine that the sheep's brain is endowed with transcendent capabilities analogous to ours. The internal complexity of the sheep's organ is indeed almost as astounding as that of the human brain, with its billions of cells, trillions of connections between cells, and ability to learn and coordinate flexible behaviors that carry us across lifespans more convoluted than the cerebral cortex. The sheep's brain bears witness to years of ovine toil, longing, passion, and caprice that are easily anthropomorphized. And that brain, removed from the rest of its body and everything the ex-sheep once felt or knew, is as powerful a memento mori as one can find.

But the sheep's brain, like ours, is also a material highly similar to other biological tissues and organs. Live brains have a jellylike consistency that can be characterized by a quantity called an *elastic modulus*, a measure of its capacity to jiggle without losing its form. The human brain has an elastic modulus of about 0.5–1.0 kilopascal (kPa), similar to that of Jell-O (1 kPa), but much lower than biological substances such as muscle or bone. Brains can also be characterized by their density. Like many other biological materials, the density of brains is close to water; given its size, an adult human brain therefore weighs about as much as a large eggplant. A typical brain is roughly 80 percent water, 10 percent fat, and 10 percent protein by weight, leaner than many meats. A quarter pound of beef brain contains 180 percent of the US recommended daily value of vitamin B_{12}, 20 percent of the niacin and vitamin C, 16 percent of the iron and copper, 41 percent of the phosphorus, and over 1,000 percent of the cholesterol—a profile somewhat resembling an egg yolk. Risk of clogged arteries aside, why not eat the brain rather than study it?

About two million years ago, near what is now the southeastern shore of Lake Victoria in Kenya, ancient hominins were doing just that. Lake Victoria itself, the largest in Africa and source of the White Nile, is less than half a million years old and was then not even a glimmer in the eye of Mother Nature. Instead, the area was an expansive prairie, roamed by our foraging forebears, who subsisted on grassland plants and the flesh of prehistoric grazing mammals that shared the terrain. Archeological findings at this site, known as Kanjera South, document the accumulation of small and midsize animal skulls at specific locations over several thousand years. The number of skulls recovered, particularly from larger animals, substantially exceeds the corresponding numbers of other bones. This indicates that animal heads were separated from the rest of their carcasses and preferentially gathered at each site. Some skulls bear the marks of human tool use, thought to reflect efforts to break open the cranial cavities and consume their contents. Brains were apparently an important part of the diet of these early people.

Why brains? In evolutionary terms, the Kanjera humans were relatively new to meat eating; carnivory in *Homo* is documented as beginning only at about 2.5 million years ago (Mya), though it is believed to have been a major factor in our subsequent development as a species. Nonhuman carnivorous families on the scene at 2 Mya had been established meat eaters for many millions of years already. The biting jaws and catching claws of the great Pleistocene cats, the giant hyenas, and the ancestral wild dogs were better adapted to slaying, flaying, and devouring their prey than anything in the contemporary hominin body plan. But early humans had advantages of their own: already the bipedal stance, the storied opposable thumb, and a nascent ability to form and apply artificial implements all conferred special benefits. If a primordial person stumbled across the carcass of a slain deer, pungent and already picked to the bone by tigers, she could raise a stone, bring it crashing down on the cranium, and break into a reservoir of unmolested edible matter. Or if she brought down an animal herself, she could pry off the head and carry it back for sharing with her clan, even if the rest of the animal was too heavy to drag. In such fashion, the hominins demonstrated their ability to carve out an ecological niche inaccessible to quadrupedal hunters. Although other carnivores competed vigorously with humans for most cuts of meat, brains may have been uniquely humankind's for the taking.

Synchronicity on a geologic time scale may explain the coincidence of early hominin brain eating and the emergence of massive, powerful brains in our genus, but the two phenomena are connected in other ways as well. Highly evolved human civilizations and their corresponding cuisines across the world have produced edible brain preparations that range from simple, everyday dishes to splendid delicacies. Celebrity chef Mario Batali brings us calf brain ravioli straight from his grandmother, needing about one hour of preparation and cooking time. Traditional forms of the hearty Mexican hominy stew called *posole* are somewhat more involved: an entire pig's head is boiled for about six hours until the meat falls off the bone. Unkosher, but perhaps appetizing all the same! Truly festive brain dishes are prepared across much of the Muslim world on the feast of sacrifice, Eid al-Adha, which celebrates Abraham's offering of his son Ishmael to God. These recipes—brain masala, brains in preserved lemon sauce, steamed lamb's head, and others—leverage the glut of ritually slaughtered animals generated on the holiday, as well as a cultural reluctance to let good food go to waste. And who could forget the highlight of Indiana Jones's Himalayan banquet on the threshold of the Temple of Doom—a dessert of chilled brains cheerfully scooped out of grimacing monkey heads? Although it is a myth that monkey brains are eaten on the Indian subcontinent, they are a bona fide, if rare, component of the proverbially catholic Chinese cuisine to the east.

Even to the hardened cultural relativist, there is something slightly savage about the idea of consuming brains as food. "It's like eating your mind!" my little girl said to me at the dinner table, a scowl on her face. Eating monkey brains seems most definitively savage because of the resemblance of monkeys to ourselves, and eating human brains is so far beyond the pale that on at least one occasion it has invited the wrath of God himself. The unhappy victims of that almighty vengeance were the Fore people of New Guinea, discovered by colonists only in the 1930s and decimated by an epidemic of *kuru*, sometimes called "laughing sickness." Kuru is a disease we now believe to be transmitted by direct contact with the brains of deceased kuru sufferers; it is closely related to mad cow disease. The Fore were susceptible to kuru because of their practice of endocannibalism—the eating of their own kind—as Carleton Gajdusek discovered in epidemiological studies that later won him a Nobel Prize. "To see whole groups of well nourished healthy young adults dancing about, with athetoid tremors which look far more hysterical than organic, is a real sight," Gajdusek wrote. "And to see them, however, regularly progress to neurological degeneration . . . and to death is another matter and cannot be shrugged off."

Fore people were surprisingly nonchalant about their cannibalism. The bodies of naturally deceased relatives were dismembered outside in the garden, and all parts were taken except the gallbladder, which was considered too bitter. The anthropologist Shirley Lindenbaum writes that brains were extracted from cleaved heads and then "squeezed into a pulp and steamed in bamboo cylinders" before eating. Fore cannibalism was not a ritual; it was a meal. The body was viewed as a source of protein and an alternative to pork in a society for which meat was scarce. The pleasure of eating dead people (as well as frogs and insects) generally went to women and children, because the more prestigious pig products were preferentially awarded to the adult males. The brain of a dead man was eaten by his sister, daughter-in-law, or maternal aunts and uncles, while the brain of a dead woman was eaten by her sister-in-law or daughter-in-law. There was no spiritual significance to this pattern, but it did closely parallel the spread of kuru along gender and kinship lines until Fore cannibalism was eliminated in the 1970s.

There are many reasons not to eat brains, from ethical objections to eating meat in general, to the sheer difficulty of the butchery, to the danger of disease; but all activities come with some difficulties and dangers. One can't help thinking that the real reason *our culture* doesn't eat brains is more closely related to the awesomeness of holding a sheep's brain in one's hand: brains are sacred to us, and it takes an exercise of willpower to think of them as just meat. Eating someone else's brain, even an animal's, is too much like eating our own brain, and eating our own brain—as my daughter asserted—is like eating our mind, and perhaps our very soul.

Some of us arrive at this conclusion through introspection. Even in the sixth century BCE, the Pythagoreans apparently avoided eating brains and hearts because of their belief that these organs were associated with the soul and its transmigration. But can we find objective data to demonstrate a modern disinclination to eat brains? Consumption of offal of all sorts, at least in Europe and the United States, has dropped precipitously since the beginning of the twentieth century, but it seems that brains in fact are particularly out of favor. A recent search of a popular online recipe database uncovered seventy-three liver recipes, twenty-eight stomach recipes, nine tongue recipes, four kidney recipes (not including beans), and two brain recipes. If we suppose somewhat crudely that the number of recipes reflects the prevalence of these ingredients in actual cooking, there appears to be a distinct bias against brains. Some of the bias may be related to "bioavailability"—a cow's brain weighs roughly a pound, compared with two to three pounds for a tongue or ten pounds for a liver—but a difference in popularity plausibly explains much of the trend. A 1990 study of food preferences surveyed from a sample set of English consumers also supports this point. The results showed that dislike for various forms of offal was ranked in ascending order from heart, kidney, tripe, tongue, and pancreas to brain. This study is notable partly because it was performed before the mad cow outbreak of the mid-1990s, so the surveyed preferences are not easily explained by health concerns related to brain eating. The participants' tendency to "identify with" brains might best explain revulsion at eating them, inferred sociologist Stephen Mennell in an interpretation of the results.

Most people lack an appetite for brains, but hunger and the brain remain closely intertwined in other ways, both literally and metaphorically. In the most concrete sense, brains are of course necessary for the perception of hunger in each of us. The cognitive basis for hunger revolves largely around a group of cells that live in a brain region called the *hypothalamus*. Some of these cells secrete a hormone called *Agouti-related peptide* (AgRP), a small protein molecule that for convoluted reasons bears the name of a winsome Mesoamerican rodent. Stimulation of AgRP release in mouse brains results in voracious feeding and an irrepressible willingness to work for food. When humans get hungry, it is possible to detect somewhat more subtle consequences. A remarkable 1945 study called the Minnesota Starvation Experiment, motivated by fears of wartime deprivation, followed the behavior and psychology of thirty-six men subjected to a semi-starvation diet in which they lost 25 percent of their body weight. "Hunger made the men obsessed with food," wrote historians David Baker and Natacha Keramidas in an account of the experiment. The subjects "would dream and fantasize about food, read and talk about food and savor the two meals a day they were given."

Our brain-starved society has also engendered a figurative hunger for brains that finds its expression in reading, talking, and fantasizing. Widespread recognition of the significance of brain function to human nature burst into the public scene during the Victorian era with the popularization of phrenology. Phrenology's founder, Franz Gall (1758–1828), claimed that he began to conceive his influential theory about the relationship between cranial features and mental capacities by observing his fellow students in primary school. Following medical studies at the Universities of Strasbourg and Vienna, Gall's social connections and occupation as physician to a Viennese lunatic asylum gave him opportunities to observe the physiognomy of patients from many walks of life. He endeavored to obtain the brains of those he had observed and tried to relate neuroanatomy to the exterior attributes he had previously recorded. Gall emerged with a set of basic tenets, chiefly that cognitive faculties are localized in discrete brain regions, that the size of these regions corresponds to the strength or power of the corresponding faculties, and that the shape of the skull reflects the underlying regional brain structure. Gall began to go public with his views in the 1790s but was censored in Austria for his secular view of human nature and ultimately induced to leave the country. Settling in Paris but traveling and lecturing widely across northern Europe, Gall became a tireless advocate for his celebrated brainchild.

Phrenology became extraordinarily influential over the subsequent decades. Its teachings penetrated the English-speaking world through the proselytizing of Gall's energetic protégé, Johann Spurzheim. His most prominent Anglophone convert, George Combe, authored an international best seller inspired by phrenology called *The Constitution of Man*. The book sold over a quarter of a million copies within thirty years of its publication in 1828; it became one of the most widely read books of the nineteenth century, vastly outstripping the scientific treatises of contemporaries such as Charles Darwin. Phrenological societies sprang up in cities across the United States and Europe. A phrenological journal was founded by the brothers Orson and Lorenzo Fowler, who also ran craniology examination parlors in New York, Boston, and Philadelphia and manufactured the iconic parcellated porcelain heads sold to this day as novelty items. Noteworthies from Abraham Lincoln to Walt Whitman underwent phrenological readings, but commercialization of phrenological ideas and techniques also divorced the discipline from its claimed scientific roots and led to denunciations of phrenology as little more than quackery. In recent years, some of Gall's theories have received a partial vindication, following the discovery of highly specialized regions in primate brains. But the deeper significance of the movement Gall and Spurzheim spawned lies in its role as the first broad-based intellectual trend that sought explanations of human behavior in the material brain.

The burgeoning of nineteenth-century interest in brains also led to the curious phenomenon of brain collecting. A menagerie of 432 brains was assembled by the neuroanatomist Paul Broca, who used his acquisitions to draw conclusions resonant with phrenological theory. The lesioned brains of some of his aphasic patients in particular led to the discovery of *Broca's area*, a frontal lobe region closely associated with language. Brains of some of Europe's brightest luminaries were posthumously harvested and examined for signatures of greatness. The brains of both Gall and Spurzheim were among those salvaged. Lord Byron's brain was one of the heaviest brains recorded, a whopping 4.9 pounds; big brains like his reinforced Eurocentric notions of racial and intellectual superiority when compared with smaller African brains such as that of the so-called Hottentot Venus, Sarah Baartman, a South African slave and performer who was dissected by the French zoologist Georges Cuvier. Cuvier's own brain weighed in at 4 pounds and Broca's at 3.3 pounds.

In a particularly striking episode, the brain of the great mathematician Carl Gauss was bequeathed in 1855 to his close friend, the Göttingen anatomist and physician Rudolf Wagner. The price of this inheritance was that Wagner had to help take it out himself. Imagine the awkwardness of performing an autopsy on an intimate acquaintance, particularly the opening of the skull and extraction of the cerebral matter itself! Wagner collaborated on the autopsy with several others, an arrangement that no doubt dissipated some of the psychic tension. One of the other participants, the prominent physician Konrad Fuchs, was also later autopsied by Rudolf Wagner. By a curious quirk of fate, Gauss's brain was accidentally exchanged with Fuchs's and the mistake went undiscovered for 150 years. Before the mix-up, Gauss's brain was weighed and found too light—about three pounds, only marginally above average for adult males and certainly insufficient to explain the prodigious cognitive capabilities of the "Prince of Mathematics." For an explanation of the genius's powers, Wagner turned instead to the deep crevices of the brain's surface, the sulci, which were a topic of interest to neuroanatomists of that day. Wagner noted that Gauss's sulci were the deepest and most convoluted he had seen. We now know, however, that these measures are only weakly related to general intelligence.

Brain collections are still actively maintained in medical establishments around the world. They now play a key role as repositories of tissue samples that facilitate the analysis of neurological diseases suffered by some of the donors. The largest brain collection is almost literally in my backyard in Belmont, Massachusetts. The Harvard Brain Bank at McLean Hospital holds more than seven thousand human brains in rooms full of stacked Rubbermaid containers and freezer arrays. Scientists and clinicians may request samples for histological or genetic studies; pieces are mailed out as dissected tissue blocks or two-dimensional vertical slices called *coronal sections*. Recruiting donors is not easy; the functioning of such resources clearly depends on the public's appreciation of the significance of brains and brain science.

Over the two hundred years since Gall, both popular and professional preoccupation with brains has grown dramatically. George H. W. Bush declared the 1990s to be the Decade of the Brain, with the stated goal to "enhance public awareness of the benefits to be derived from brain research." Soon after this decade had run its course, the US National Institutes of Health (NIH, the world's biggest sponsor of medical research) announced its 2004 "Blueprint for Neuroscience Research," an effort to boost neurobiology research and technology through a series of focused objectives and "grand challenges" to scientific investigators. In 2013, both the US federal government and the European Union announced ambitious further endeavors to promote and integrate future brain research. Ever-increasing participation in brain science is reflected in the attendance statistics at the monolithic Society for Neuroscience annual conference, which saw peak attendance of about six thousand people in the 1970s, fourteen thousand in the '80s, twenty-six thousand in the '90s, and thirty-five thousand in 2000. The neuroscience conference now has a population greater than most towns in America.

Consumption of brain-related literature has also followed a trajectory of rapid expansion. The number of print books listed on Amazon with key word "brain" has approximately doubled with each decade since the 1970s, an exponential growth pattern similar to the famous Moore's Law, which predicts the doubling of computer processing power at regular intervals. Of the 5,070 "brain" books listed on Amazon in 2014, 164 books were published in the 1970s, 470 in the '80s, 983 in the '90s, 1,676 in the '00s, and more than 1,500 published in the first half of the teens—on track to continue the doubling trend into the current decade. Over the same time span, entries associated with key word "brain" or "neuron" in the US National Library of Medicine, the definitive index of life science publications, have steadily grown from 13,000 per year in 1970 to more than 60,000 per year since 2010.

On undergraduate college campuses throughout America, similar trends are apparent. At most schools, the closest major to neuroscience is psychology, a framing subject that includes behavioral, cognitive, and biological components. Psychology is reported to be the second most popular major at US colleges, surpassed only by business. The number of students graduating with degrees in psychology has increased vastly from a total of about 38,000 in 1970 to over 100,000 per year in recent years. As a child, I was amazed to learn that a large concert hall near my mother's office at Cornell University doubled as a classroom for the 1,600 students of Psych 101, the introductory psychology class there.

Comparable mega-classes in psychology are common across the country and for countless pupils provide an opportunity to begin investigating the innards of their minds and brains.

Trends in education and the media have made us more aware than ever of the importance of brains in our lives. Our appetite for literature and lectures about brains is only a component of this trend. At a more intimate level, most of us have had friends or relatives affected by brain diseases such as Alzheimer's or Parkinson's. We may also have personal reasons to know about the danger of concussions and head injuries, or about drug abuse and its effects on the brain. Findings backed by solid science but recognized in the past only by specialists have slowly been entering the popular consciousness. Through this exposure, we have learned some of the ways that our brains are important for perception and cognition, and we now have falsifiable hypotheses about how these phenomena might work. The demise of phrenology notwithstanding, we have seen that different brain regions really can do different things. Our brains can also undergo changes, store memories, help make decisions, and commit errors. Basic neuroscience research has even given us insights into the specific molecular and cellular factors involved when our brains change, remember, decide, and err.

But have we ourselves been changed by what we know about the brain? If neuroscience teaches us that our minds are based on biological processes, shouldn't our attitudes and practices be radically affected? Why haven't our notions of personal responsibility and individual identity fundamentally changed? Why does our society still punish and reward people in virtually the same ways that it did a hundred years ago? Why do we continue to stigmatize mental disorders more than kidney disease or pneumonia? Why do we feel differently about medicines and technologies that act on the brain compared with those that act on the rest of the body? Some might argue that our neuroscientific understanding of core mental processes is still too rudimentary to make a difference in addressing real-world problems. But our society didn't need microscopic understanding of infectious agents in order to give up bloodletting as the remedy to diseases in the nineteenth century. Similarly, most college-educated people don't need a full description of the factors involved in climate change, macroeconomic theory, or tribalism in Afghanistan in order to grasp some of the basics and think about their policy implications. So if neuroscience hasn't changed our worldview in important ways already, what is getting in the way?

One answer is that despite growing awareness of brain science, most of us continue to live our lives in an extraordinary level of denial about the biological nature of our minds and selves. We routinely distinguish *mental* from *physical* worlds both in conversation and in analysis. Even if we accept intellectually that cognition arises from physical phenomena in and around our brains, we operationally wall this fact off from our conscious actions and thoughts. There is no growling, congestion, or tingling to disrupt our daily reveries and remind us of the brain's quirky presence. For the most part, the function of the human brain therefore remains abstract, unfathomable, and remote. Like events in far-off countries, neurobiological discoveries make for engrossing reading or research but still leave us largely untouched. To be changed by neuroscience, we need to get more personal with the brain, and to get personal we need to lose some of the exaggerated sense of wonder that distances us from the organ of our minds.

Pictures can begin to show us how fascination with the brain leads to unrealistic views of its role. When brains appear in magazines or animations, they are surreal, free-floating forms, often blue and iridescent like Luke Skywalker's first lightsaber, sparkling with occult energy (see Figure 1). The brain I remember best from my youth lived in a tank of green slime and pulsated with light; it belonged to the *Doctor Who* television series villain Morbius, a megalomaniacal mass murderer who had met with an unfortunate accident somewhere along the line. In scientific images, brains are most likely to be speckled with fluorescent colors or flashing bright spots of red and yellow to denote patterns of activity in scans. Even on the covers of neurology textbooks, brains are apt to be glistening, glowing, or ghostly as X-rays. These images at once project power and enigma, like the chryselephantine idols of the ancients. They evoke depictions of the Holy Spirit as a radiant bird in Renaissance paintings or the glowing halos that emanate from gods and saints in religious art from around the world.

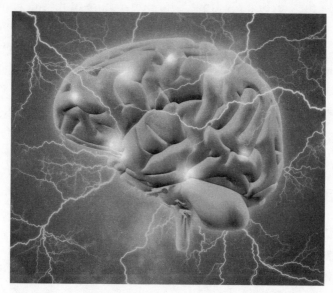

FIGURE 1. A typical mystified brain. *(Licensed from Adobe Stock)*

Images often express feelings that words or conscious thoughts convey with less fidelity. Picasso was known to have increasingly distorted the unattractive features of his mistresses in portraits as he fell out of love with them. "It must be painful for a girl to see in a painting that she is on the way out," the artist once remarked. The psychologist Carl Jung discerned evidence of unconscious mental representations in imagery ranging from ancient religious artifacts to the hallucinations of his schizophrenic patients. Some images of glowing brains in fact bear a strong resemblance to one of Jung's archetypal forms, the "solar phallus," which he presented as a libidinous semi-religious image recurring spontaneously throughout history. Below the bright hemispheres of the cerebral cortex protrudes the priapic medulla oblongata, the brain's regulator of vital primitive functions such as breathing and heart rate. Surely Jung would have enjoyed the resemblance.

The supernatural iconography of the brain both reflects and reinforces a romanticized view of what goes on between the ears—a mystique of the brain, what it does, and what it makes us do. This *cerebral mystique* drives many of us to see the brain as the essence of our humanity, to reduce our problems to its problems, and to study it rather than eat it. Like other mystiques, the cerebral mystique connotes a sense of mystery and magic, a charm and charisma that distinguishes the brain from merely academic concerns. Lots of things can become interests or even obsessions (cooking, stamp collecting, Dungeons & Dragons), but few of these possess the *je ne sais quoi* that makes for a true mystique. Scientific problems don't usually engender mystiques. Even some of the most compelling topics of the day—the causes of cancer, the properties of newly discovered materials, machine-learning algorithms—may inspire intense commitment but fail to arouse the allure that for many people surrounds the brain. Mystiques develop most powerfully around those scientific fields that engage with essential aspects of existence, such as the origin of the universe or the nature of consciousness.

Mystiques are invigorating, but they are also impediments to enlightenment and progress. The "feminine mystique" of Betty Friedan's revolutionary 1963 book referred to an entrenched set of constraining attitudes about the proper place of women in society. Friedan blamed this mystique for women's reluctant decision to shelve their ambitions in order to take on traditional womanly roles in the home. Mystiques associated with faraway places have been rife and consequential. For centuries across Europe, the mystique of the East in particular evoked florid literary expressions that have come to be known as Orientalism. This cultural movement, with its objectification of Eastern people and their traditions, is now cited as one of the pillars of colonialism. Some would say that science as a whole possesses a mystique that has at times been abused. The mystique of science has been borrowed by fields that do not deal with the natural world or that lack the deterministic characteristics commonly associated with the natural sciences. The prestige of scientific objectivity was most insidiously harnessed to justify the racist theories that contributed to European imperialism and the atrocities of World War II.

The cerebral mystique is a similarly powerful illusion about the exceptional qualities of our brains and thus about ourselves as individuals. In its sway, we escape the inevitable implications of having a biologically based mind and idealize the mind's chief organ in ways we will examine in this book. We echo the spirituality of the past when we conceive of the brain as an omnipotent structure that encapsulates everything important about our personalities, intellect, and will. In effect, the cerebral mystique results in a psychological *transference* of old beliefs regarding the soul to new attitudes about the brain. Freud wrote that transference could be alleviated first by making the patient conscious of its effect. In the rest of Part 1 of this book, we will place ourselves on the therapist's couch by examining and deconstructing several manifestations of the neurofantasy that keeps us in denial about our organic selves.

two

HUMOR ME

A DEFINING CHARACTERISTIC OF THE CEREBRAL MYSTIQUE is the artificial distinction it draws between brain and body. In this chapter we will see how abiotic depictions of the brain, and in particular the pervasive analogy of the brain to a computer, promote this distinction. The true brain is a grimy affair, swamped with fluids, chemicals, and glue-like cells called *glia*. The centerpiece of our biological mind is more like our other organs than a man-made device, but the ways we think and talk about it often misrepresent its true nature.

Standing astride a pilaster on the facade of Number 6 Burlington Gardens in London's Mayfair district is a statue that conjures up the long history of such misconceptions. It is the likeness of Claudius Galenus of Pergamon, more commonly known as Galen, possibly the most influential figure in the history of medicine. His sneer of cold command conveys the haughtiness of a man who learned his trade at the gladiator's arena, ministered to four Roman emperors, and reigned as an unchallenged oracle of medical truth for over a thousand years. In Galen's stony hands rests a skull, symbolizing the biological principles he revealed in public dissections before audiences of Roman aristocrats and academics. Galen's place in the pantheon of great intellectuals reflects his discoveries, as well as the endurance of his copious writings, over three million words by some counts, which over centuries were copied, amplified, and elaborated like scripture by Arabic and European scholars. Far from his birthplace in Asia Minor, the Galen of Burlington Gardens is flanked by statues of similarly iconic luminaries dear to Victorian scientific culture. He is of course a fabrication—no likeness of the real Galen survives from his own time.

Galen's investigations contributed significantly to the triumph of brain-centered views of cognition. Although Hippocrates of Kos, writing four hundred years before Galen's time, had already proclaimed the brain as the seat of reason, sensation, and emotion, Roman contemporaries of Galen maintained the Aristotelian cardiocentric view that the heart and vascular system controlled the body, including the brain. To Galen, the heart and vasculature occupied the crucial but subsidiary role of supplying "vital spirits" that energize the body. Galen's vote for the brain followed largely from observation of the relationships of gladiators' wounds to deficits they displayed in action, an incisive data source blissfully unavailable to later scientists.

Galen also performed careful dissections, raising this approach to an art. His dissections were performed exclusively on animals; human bodies were considered sacred (at least outside the arena) and not to be defiled by experimentation, even after death. Galen traced the peripheral nerves of his subjects to their origins at the base of the brain, providing evidence that the brain was uniquely capable of controlling the body. One famous experiment involved severing one of these fibers, the laryngeal nerve, in the head of a live pig, an operation that rendered the pig mute. Galen probably sent his slaves to procure carcasses and body parts from the local markets. Butchered heads were widely available at the time, no doubt destined for the tables of the well-to-do. The doctor carved them up to reveal notable features of intracranial anatomy. He took particular interest in structures he thought to be interfaces between the vasculature and the brain. Galen viewed these structures as critical for the conversion of vital spirits into "animal spirits," the fluid essence to which he attributed consciousness and mental activity. Candidate interfaces included the linings of the ventricles, the fluid-filled cavities common to vertebrate brains, as well as a curious weblike structure of interconnected blood vessels so singular in Galen's anatomical investigations as to merit the appellation *rete mirabile,* or "wondrous net."

The *rete* figured prominently in Galen's writings about the brain. It was in effect a biological locus of ensoulment, and reverence for the importance of this structure was passed down with Galen's writings as received truth for hundreds of years. Like the statue at Burlington Gardens however, the *rete mirabile* was a mirage. Renaissance anatomists discovered that the formation occurs only in animals but not in people. In his monumental *De Humani Corporis Fabrica* (1543), the pioneering anatomist Andreas Vesalius wrote confidently that the blood vessels at the base of the human brain "quite fail to produce such a *plexus reticularis* as that which Galen recounts." Galen's extrapolations from animal dissections were indeed erroneous, his conclusions skewed by the cultural taboos of his time. Yet as a symbol of the brain's mysterious qualities, the *rete mirabile* continued to appeal long after it was discredited scientifically. A hundred years after Vesalius, Galen's obsession inspired the English poet John Dryden to write, "Or is it fortune's work, that in your head / The curious net, that is for fancies spread / Lets through its meshes every meaner thought / While rich ideas there are only caught?"

The story of Galen's *rete* shows us that salient but arbitrary or even mistaken features of the brain can be singled out for special attention because they mesh with the culture of the time. In Galen's day, pride of place went to the *rete mirabile* and its part in a theory of the human mind that was governed by spirits. As we shall see in this chapter, the importance now ascribed to neuroelectricity and its role in computational views of brain function occupies a similar position in our era. I will argue that our cerebral mystique is upheld by contemporary images of the brain as a machine. I will also present a more organic, alternative picture of brain function that tends to demystify the brain, and that also bears curious resemblance to the ancient theory of spirits.

Like other wonders of nature, the brain and mind have always been popular subjects of poetic conceit. Long before Dryden, Plato wrote that the mind was a chariot steered by reason but pulled by the passions. Anchored by considerably deeper biological insight in 1940, the groundbreaking neurophysiologist Charles Sherrington described the brain as "an enchanted loom where millions of flashing shuttles weave a dissolving pattern, always a meaningful pattern though never an abiding one; a shifting harmony of subpatterns." The loom metaphor has found its way into the titles of several books and even has its own Wikipedia page. Sherrington's fibrous motif evokes Galen's net; his musical reference also resonates with the imagery of other writers, who analogized the brain to a piano or a phonograph, both of which mimic the brain's capacity to emit a large repertoire of complex but chronologically organized output sequences. In his book *The Engines of the Human Body* (1920), the anthropologist Arthur Keith laid out the more prosaic comparison to an automated telephone switchboard, conceptualizing the brain's ability to connect diverse sensory inputs and behavioral outputs.

The most popular analogy for the mind today is the computer, and for good reason—like our minds, modern computers are capable of inscrutable feats of intellect. Critics have objected to the notion that human consciousness and understanding can be reduced to the soulless digit crunching performed by CPUs. To the extent that the analogy between minds and computers ignores or trivializes consciousness, it demeans what we consider most special about ourselves. The computational view of the mind took off at a time when human minds so clearly outranked computers that the insult carried a bit more bite than it does today. The situation is almost reversed now: we associate computers with a combination of arithmetic acumen, memory capacity, and accuracy that our own minds certainly cannot equal.

Most scientists and philosophers accept the analogy between minds and computers and actively or passively incorporate it into their professional creeds. Given the close association between mind and brain, a computational view of the brain itself is likewise widespread. The portrayal of brain as computer permeates our culture. One of the most memorable episodes from the original *Star Trek* television series begins when an alien steals Mr. Spock's brain and installs it at the core of a giant computer, where it controls life support systems throughout an entire planet. The robots of science fiction generally have brainlike computers or computerlike brains in their heads, ranging from the positronic brains of Isaac Asimov's *I, Robot* to the dysfunctional brain that occupies the oversized cranium of Marvin the paranoid android, in the 2005 film version of *The Hitchhiker's Guide to the Galaxy*. In contrast, many of the real-life robots sponsored by the US Defense Advanced Research Projects Agency (DARPA) wear their processors in their chests or even distributed throughout their bodies, where they are somewhat less brainlike but better protected against mishaps. Popular science magazines are full of the brain-computer analogy, comparing and contrasting brains with actual computers in terms of speed and efficiency.

But what is the "meat" in a computational view of the brain—does the comparison really help us understand anything? Fingers are like chopsticks. Fists are like hammers. Eyes are like cameras. Mouths and ears are like telephones. These analogies are not worth dwelling on because they are too obvious. The tool in each pairing is an object designed to do a thing that we humans have evolved to do but wish we could do better, or at least slightly differently—that's why we made the tools. At some point we also decided we wanted to multiply numbers bigger or faster than we could manipulate easily in our heads, so we built tools to accomplish this. Similar tools turned out to be useful for various other things we also do with our brains: remembering things, solving equations, recognizing voices, driving cars, guiding missiles. Brains are like computers because computers were designed to do things our brains do, only better.

Brains are enough like computers that physical analogies between brains and computers have been proposed since the earliest days of the digital age, when John von Neumann, the mathematician and computing innovator, wrote *The Computer and the Brain* in 1957. Von Neumann argued that the mathematical operations and design principles implemented in digital machines might be similar to phenomena in the brain. Some of the similarities that prompted von Neumann's comparison are well-known. Both computers and brains are noted for their dependence on electricity. Neuroelectricity can be detected remotely using electrodes placed outside brain cells and even outside the head, making electrical activity a particularly salient hallmark of brain function. If you've ever had an electroencephalography (EEG) test, you've seen this phenomenon in

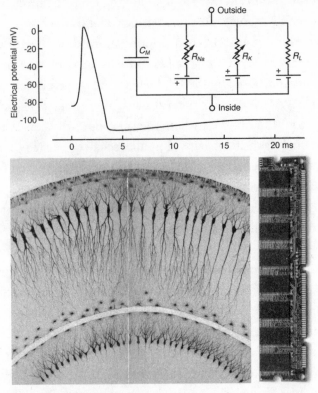

FIGURE 2. Electronic and computational analogies to brain function: (top) trans-membrane voltage versus time during an action potential, with inset showing a circuit model that predicts neural membrane potentials, labeled according to the conventions of electronics, after work by A. L. Hodgkin and A. F. Huxley; (bottom left) neural structure of the hippocampus, as illustrated by the famous neuroanatomist Camillo Golgi; (bottom right) memory circuit board from a modern computer. (*Licensed from Adobe Stock*)

action when tiny wires were pasted to your scalp (or perhaps attached via a cap) in order to permit electrical recording of your brain activity. Such procedures help doctors detect signs of epilepsy, migraines, and other abnormalities.

The brain's electrical signals arise from tiny voltage differences across the membranes that surround neurons, like the differences between terminals on a battery (see Figure 2). Unlike batteries, transmembrane voltages (known as membrane potentials) fluctuate dynamically in time, resulting from the flow of electrically charged molecules called *ions* across the cell membrane. If the voltage across a neural membrane fluctuates by more than about twenty millivolts from the cell's resting level, a much larger voltage spike called an *action potential* can occur. During an action potential, a neuron's voltage changes by about a hundred millivolts and returns to the baseline in the space of a few milliseconds, as ions zip back and forth through little channels in the membrane. When a neuron displays such flashes of electrical energy, we say it is "firing." Action potentials spread spatially along neural fibers at speeds faster than a sprinting cheetah and are essential to how distant parts of the brain can interact rapidly enough to mediate perception and cognition.

Most neurons fire action potentials at frequencies ranging from a few per second up to about one hundred per second. In these respects, neuronal action potentials resemble the electrical impulses that make our modems and routers flash and allow our computers and other digital devices to calculate and communicate with each other. Measurements of such electrophysiological activity are the mainstay of experimental neuroscience, and electrical signaling is often thought of as the language brain cells use to talk to each other—the lingua franca of the brain.

Brains contain circuits somewhat analogous to the integrated circuits in computer chips. Neural circuits are made up of ensembles of neurons that connect to one another via *synapses*. Many neuroscientists regard synapses as the most fundamental units in neural circuitry because they can modulate neural signals as they pass from cell to cell. In this respect, synapses are like transistors, the elementary building blocks of computer circuitry that get turned on and off and regulate the flow of electric currents in digital processing. The human brain contains many billions of neurons and trillions of synapses, well over the number of transistors in a typical personal computer today. Synapses generally conduct signals in one predominant direction, from a *pre*synaptic neuron to a *post*synaptic neuron, which lie on opposite sides of each synapse. Chemicals called *neurotransmitters*, released by presynaptic cells, are the most common vehicle for this communication. Different types of synapses, often distinguished by which neurotransmitter they use, allow presynaptic cells to increase, decrease, or more subtly affect the rate of action potential firing in the postsynaptic cell. This is somewhat analogous to how your foot pressing on the pedals of a car produces different results depending on which pedal you push and what gear the car is in.

The structure of neural tissue itself sometimes resembles electronic circuitry. In many regions of the brain, neurons and their synaptic co-tacts are organized into stereotyped patterns of local connectivity, remi-niscent of the regular arrangements of electronic components that make up microchips or circuit boards. For example, the cerebral cortex, the convoluted rind that makes up the bulk of human brains, is structured in layers running parallel to the brain surface, resembling the rows of chips on a computer's memory card (see Figure 2).

Neural circuits also do things that electronic circuits in digital pro-cessors are designed to do. At the simplest level, individual neurons "compute" addition and subtraction by combining inputs from presyn-aptic cells. Roughly speaking, a postsynaptic neuron's output represents the sum of all inputs that increase its firing rate minus the sum of all inputs that decrease its activity. This elementary neural arithmetic acts as a building block for many brain functions. In the mammalian visual system, for instance, signals from presynaptic neurons that respond to light in different parts of the retina add up when these cells converge onto individual postsynaptic cells. Responses to progressively more so-phisticated light patterns can be built by combining such computations over multiple stages, each involving another level of cells that gets input from the previous level.

The complexity of neural calculations eventually extends to concepts from college-level mathematics. Neural circuits perform calculus—a mainstay of freshman-year education—whenever they help keep track of how something in the world is changing or accumulating in time. When you fix your gaze on something while moving your body or head, you are using a form of this neural calculus to keep track of your accumu-lated movements; you use the data to adjust your eyes just enough in the opposite direction so that the direction of your gaze doesn't change as you move. Scientists have found a group of thirty to sixty neurons in the brain of a goldfish that seems to accomplish this computation. A dif-ferent form of neural calculus is required for detecting moving objects in the visual system of a fly. To make this possible, small groups of neurons in the fly's retina compare input from neighboring points in space. These little neural circuits signal the presence of motion if visual input at one point arrives before input to the second point, sort of like the way you could infer motion of a subway train by considering its arrival times at adjacent stations, even if you could not directly see the train moving.

Neuroscientists speak of circuits that perform functions much more complicated than calculus as well—processes that include object recognition, decision making, and consciousness itself. Even if entire neural networks that perform these operations have not yet been mapped, neuronal hallmarks of complicated computations have been discovered by comparing the action potential firing rates of neurons to performance in behavioral tasks. One example comes from a classic set of experiments on the neural basis of learning performed by Wolfram Schultz at Cambridge University, using electrode recordings from monkey brains. Schultz's group studied a task in which the monkeys learned to associate a specific visual stimulus with a subsequent juice reward—a form of the same experiment Pavlov conducted with his dogs. In the monkeys, firing of dopamine-containing neurons in a brain region called the *ventral tegmental area* initially accompanied delivery of the juice. As animals repeatedly experienced the visual stimulus followed by the juice, however, the dopamine neurons eventually began to fire when the stimulus appeared before the juice. This showed that these neurons had come to "predict" the juice reward that followed each stimulus. Remarkably, the behavior of dopamine neurons in this task also closely paralleled part of a computational algorithm in the field of machine learning. The similarity between the abstract machine-learning method and the actual biological signals suggests that the monkeys' brains might be using neural circuits to implement an algorithm similar to the computer's.

In a further parallel between electrical engineering and the activity attributed to the brain, neuronal firing rates are often said to encode *information*, in reference to a theory Claude Shannon developed in the 1940s to describe the reliability of communication in electronic systems like radios or telephones. Shannon's information theory is used routinely in engineering and computer science to measure the reliability with which inputs are related to outputs. We implicitly brush up against information theory when we compress megapixel camera images into kilobyte jpeg images without losing detail, or when we transfer files over the ethernet cables in our homes or offices. To make these tasks work well, engineers had to think about how effectively the compressed data in our digital photographs can be retrieved or how accurately and quickly the signal transmitted over cables can be understood or "decoded" at the other end of each upload or download. Such problems are closely related to the questions of how data are maintained in biological memory and how the timing of action potentials communicates sensory information along nerve fibers to the brain. The mathematical formalisms of information theory and of signal processing more generally can be tremendously useful for quantitative interpretation of neural functions.

When we think of the brain as an electronic device, it seems entirely natural to analyze brain data using engineering approaches such as information theory or machine-learning models. In some cases, the computational analogy of the brain drives researchers even further—to imagine that parts of the brain correspond to gross features of a computer. In a 2010 book, the neuroscientists Randy Gallistel and Adam King argued that the brain must possess a read-write memory storage device similar to that of a prototypical computer, the Turing machine. The Turing machine processes data by writing and reading zeros and ones from a piece of tape; the reading and writing operations proceed according to a set of rules in the machine (a "program"), and the tape constitutes the machine's memory, analogous to the disks or solid-state memory chips used in modern PCs. If efficient computers universally depend on such read-write memory mechanisms, Gallistel and King reason, then the brain should too. The authors thus challenge the contemporary dogma that the basis of biological memory lies in changing synaptic connections between neurons, which are difficult to relate to Turing-style memory; they insist that this synaptic mechanism is too slow and inflexible, despite the formidable experimental evidence in its favor. Although Gallistel and King's hypothesis is not widely accepted, it nevertheless offers a remarkable example of how the analogy between brains and computers can take precedence over theories derived from experimental observation. In looking from brain to computer and back from computer to brain, it can be difficult to tell which is the inspiration for which.

The association of brains with computers sometimes seems to take on a spiritual flavor. John von Neumann's own early efforts to synthesize computer science and neurobiology apparently coincided with his rediscovery of Catholicism, shortly before his death of pancreatic cancer in 1957. There is little evidence that religion was at all important to von Neumann throughout much of his life, although he had undergone baptism in 1930 on the eve of his first marriage. It is a cliché that people find God on their deathbed—a kind of last-minute insurance for the soul—and at first it might seem dissonant to think at the same time about recasting the material basis of the soul itself into the language of machines. From another angle, these views are easy to reconcile, however, because equating the organic mind to an inorganic mechanism might offer hope of a secular immortality—if not for ourselves, then for our species. If we are our brains, and our brains are isomorphic to devices we could build, then we can also imagine them being repaired, remade, cloned, propagated, sent through space, or stored for an eternity in solid-state dormancy before being awakened when the time is right. In identifying our brains with computers, we also tacitly deny the messy, mortal confusion of our true physical selves and replace it with an ideal not born of flesh.

A substantial cohort of eminent physical scientists in their later lives joined von Neumann by also speculating about abstract or mechanical origins of cognition. With the wave equation almost twenty years behind him and his renowned cat nine years out of the bag, Erwin Schrödinger postulated a universal consciousness embodied in the statistical motion of atoms and molecules. His theory is far removed from von Neumann's computer analogy but likewise presents mental processes as fundamentally abiotic. Another case in point is that of Roger Penrose, the eminent cosmologist whose contributions to the understanding of black holes are overshadowed in some circles by his commentaries on consciousness. Penrose explicitly rejects the suggestion that a computer could emulate human minds but instead seeks a basis for free will in the esoteric principles of quantum physics. Like the computer analogy, Penrose's quantum view of the mind seems rooted more in physics than physiology and in equations more than experiments. The biophysicist Francis Crick turned to neuroscience after codiscovering the structure of DNA; his influence lingers powerfully in his injunction that researchers should seek correlates of consciousness in the electrical activity of large neuronal ensembles. But even Crick's ruthlessly materialist and biologically anchored view of the brain focuses almost entirely on computational and electrophysiological aspects of brain function that most differentiate the brain from the rest of the body.

Although each of these perspectives differs dramatically from the others, they share a tendency to minimize the organic aspects of brains and minds and emphasize inorganic qualities that relate most distantly to other biological entities. In effect, they set up a *brain-body distinction* that parallels the age-old metaphysical distinction between *mind* and body, traditionally referred to as *mind-body dualism*. Through this distinction, the brain takes the place of the mind, and thus becomes analogous to an immaterial entity humankind has struggled for millennia to explain.

The tendency to draw a distinction between the brain and the rest of the body is a phenomenon I will call *scientific dualism*, because it parallels mind-body dualism but draws strength from strands of scientific thought and coexists with scientific worldviews. Scientific dualism is one of the most ubiquitous realizations of the cerebral mystique, and we shall see it in many forms throughout this book. It is the powerful cultural vestige of a philosophy most commonly associated with René Descartes, a seventeenth-century scholar and adventurer who argued that mind and body are made of separate substances that interact to actuate living beings. In Descartes's depiction, the mind or soul (he made no distinction) interacts with the body through part of the brain, though Descartes was never able to explain the mechanics of how this interaction could take place. Related forms of dualism in which the soul departs the body upon death, submits to divine judgment, and sometimes finds a new body are almost universally present among the religions of the world.

Dualism is an operating principle most of us use at least implicitly in daily life. Even outside our places of worship, and even if we are not religious, we speak of the mind and spirit in ways that distinguish it from the body. We say that so-and-so has lost his mind or that what's-her-name lacks spirit. The ego and id of Freudian psychoanalysis, now fixtures of folk psychology, lead dualism-sanctioned lives of their own: "My ego tells me to do this; my id tells me to do that." And our actions also reflect dualism. For example, a white-collar workaholic who fails to connect the importance of a sound mind to the need for a sound body may be in for an early heart attack, and could well suffer diminished productivity even before the sad corporeal end comes. In other instances, we might fear judgment about mental transgressions that could never possibly be witnessed by other people—Jesus referred to this as sinning "in one's heart" (Matthew 5:28), but atheists probably know the feeling just as well. Our anxiety here is a manifestation of dualism because we suppose at least subconsciously that the mind can be accessed separately from the body, perhaps even after we die.

In traditional dualist perspectives like Descartes's, the mind or soul is like the invisible operator of a remote-controlled body. In scientific dualism, on the other hand, the operator is not an incorporeal entity but rather a material brain, which lives within the body but otherwise fulfills the same mysterious role. Unlike the dualisms of religion and philosophy, scientific dualism is rarely a consciously held opinion or an openly professed point of view. Few scientifically informed people really believe that the brain and body are materially separable, but they might nevertheless treat the brain and body separately in thought, rhetoric, and even practice. Through scientific dualism, some of the cherished attitudes about the disembodied soul can thus persist without any conviction that the soul or mind is truly incorporeal. In this respect, scientific dualism mirrors the instinctive morality of many atheists or the tacit sexism and racism that pervade even the most enlightened corners of our postmodern society. In each of these examples, old-fashioned habits of thought outlive overt adherence to the religious or social doctrines that originally spawned them.

As with other prejudices, scientific dualism can sometimes be expressed explicitly. Take for instance the Xbox video game Body and Brain Connection, which "integrate[s] cerebral and physical challenges for the optimal gaming experience." Despite the talk of integration, the language used here treats brain and body as discrete units with functions that complement each other but do not overlap. Less explicit instances of scientific dualism arise when scientists like von Neumann, Schrödinger, Penrose, and Crick conjure up abiotic images of brains that lack the wet and squishy qualities that characterize other organs and tissues. These authors do not draw bright-line boundaries between brain and body, but their writing still implies that the brain is special in its makeup or modes of action. In each instance, scientific dualism provides a mechanism for keeping our minds sacred—distinguishing the functions and processes of the brain from those of mundane bodily processes like digestion or cancer, and perhaps even guarding our brains from being eaten. We shall see, however, that more organic views of brain physiology were once common and are being increasingly resurrected by recent science.

On a February morning in 1685, King Charles II of England emerged from his private chamber to undergo his daily toilet. His face looked ghastly, and he spoke to his acolytes with slurred speech, his mind apparently wandering. As he was being shaved, the king's complexion suddenly turned purple, and his eyes rolled back into his head. He tried to stand and instead slouched into the arms of one of his attendants. He was laid out on a bed, and a doctor stepped forward with a penknife to lance a vein and draw blood. Hot irons were applied to the monarch's head, and he was force-fed "a fearsome decoction extracted from human skulls." The king regained consciousness and spoke again, but appeared to be in terrible pain. A team of fourteen physicians waited on him and continued to draw blood—some twenty-four ounces in total—but it became apparent that they could not save him. His highness passed away four days later.

Although rumors about poison circulated at the time, the more widespread belief was that Charles II had died following an episode of apoplexy, what we now call a stroke, in which the blood vessels of the brain become blocked or broken. Strokes affect tens of millions of people each year worldwide and are still a leading cause of neurological injury and death. We have now developed treatments that reduce the risks of strokes and help protect the brain when they occur. To the seventeenth-century mind, however, brain ailments such as apoplexy, as well as diseases affecting all aspects of the body, were brought on by imbalances among bodily fluids called *humors*. An excess of blood, one of the four humors along with black bile, yellow bile, and phlegm, was thought to cause the apoplexy. Bloodletting was supposed to relieve the excess and help the patient accordingly.

Many of us remember being taught to laugh at humorism in school, and the brain in particular is difficult to imagine as a soup of bodily fluids. Current *neuronocentric* views about brain function in cognition are

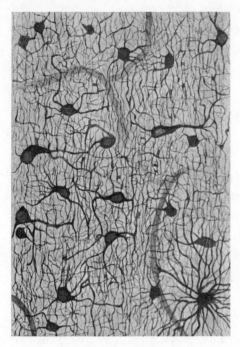

FIGURE 3. A 1928 hand-drawn illustration of glial cells by the Spanish neuroscientist Pío Del Río Hortega, also showing blood vessels (thick light gray curves), in the cerebellum of a cat's brain.

most concerned with the roles of neurons and neuroelectricity, features that lend themselves best to computational analogies and that seem inherently dry and machinelike. But although computers are known to react poorly to liquids (try spilling a cup of coffee on your laptop), the brain is actually rich in fluids that participate intimately in neurobiology. A fifth of the brain's volume consists of fluid-filled cavities and interstices. About half of this is occupied by blood, and the other half by cerebrospinal fluid (CSF), a clear substance produced by the linings of the brain's cavernous ventricles in a process strikingly resembling Galen's proposed generation of animal from vital spirits. CSF fills the ventricles and exchanges rapidly with extracellular inlets that directly contact all the brain's cells, bathing them with a mix of ions, nutrients, and molecules related to brain signaling. The cells of the brain themselves, about 80 percent by volume, are also filled with intracellular fluids, which hold the DNA and other biomolecules and metabolites that make cells work.

Perhaps more surprisingly, only at most half of brain cells are actually the charismatic, electrically active neurons that steal most neuroscientists' attention. The less noticed brain cells are the glia, smaller nonspiking cells that do not form long-range connections reminiscent of electrical wiring (see Figure 3). These cells were historically thought to play only a literally supporting role in the brain—the term *glia* derives from the Greek word for glue, another fluid—but in the cerebral cortex they outnumber neurons by up to a factor of ten to one. A conception of the brain that doesn't include a role for glia is like a brick wall built without mortar.

Oddly enough, it is precisely the nonneuronal components of brain anatomy that are often directly implicated in many of the best-known brain diseases. One of the most prevalent and pernicious brain cancers, glioblastoma multiforme, arises from uncontrolled proliferation of glial cells; the cancer then results in brain fluid pressure buildup that in most cases is the ultimate cause of death. This terrible disease is the one that killed Senator Ted Kennedy of Massachusetts in 2009. Disruptions to fluid exchange between blood vessels and surrounding brain tissue are closely associated with stroke, multiple sclerosis, concussion, and Alzheimer's disease. Many of these conditions specifically affect blood flow or the integrity of the blood-brain barrier, a network of tightly connected cells that surrounds blood vessels and regulates transport of chemicals between the blood and the brain.

Is the thinking brain really distinct from the brain that underlies neurological disease? Research now suggests that the brain's glue and fluids, previously thought to be bystanders, are in fact deeply engaged in many aspects of function. One of the striking revelations of recent years has been the discovery that glia undergo signaling processes similar to neurons. By analyzing microscopic-scale videos of neurons and glia, researchers have shown that glia respond to some of the same stimuli that neurons do. Several neurotransmitters evoke calcium ion fluctuations in glia, a phenomenon also observed in neurons, where such dynamics are closely related to electrical activity. Calcium fluctuations in a type of glial cell called an *astrocyte* are correlated with the electrical signals of nearby neurons. My MIT colleague Mriganka Sur and coworkers showed that astrocytes in the visual cortex of ferrets are even more responsive than neurons to some visual features.

Blood flow patterns in the brain are also closely correlated with neuronal activity. When regions of the brain become activated, local blood vessels dilate and blood flow increases in a coordinated phenomenon called *functional hyperemia*. Discovery of functional hyperemia is attributed to the nineteenth-century Italian physiologist Angelo Mosso. Using an oversize stethoscope-like device called a plethysmograph, he monitored pulsation of blood volume in the head noninvasively through the fontanelles of infants and in adults who had suffered injuries that breached their skulls. Mosso's best-known subject was a farmer named Bertino, whose cerebral pulsation accelerated when the local church bells rang, when his name was called, or when his mind was engaged by various tasks. These experiments were forerunners of modern brain-scanning techniques, which use positron emission tomography (PET) and magnetic resonance imaging (MRI) in place of the plethysmograph to map blood flow changes in three dimensions.

That glia and blood vessels respond to many of the same stimuli that activate neurons highlights the multifarious nature of brain tissue—neurons have housemates—but this fact does not prove that nonneuronal elements have more than a supporting role. A neuronocentric, computational view of brain function might suppose that glia and vessels are analogous to the power supply and cooling fans that keep the electronics running; they face demands that rise and fall depending on the CPU's workload, but they do not compute anything themselves. If this description were accurate, then stimulating glia or vasculature independent of neurons would have negligible effects on the activity of other neurons—but recent results contradict this premise.

Some evidence suggests, for instance, that blood flow changes can influence neural activity in addition to responding to it. Certain drugs that act on enzymes in blood vessels appear to alter neural electrical activity indirectly, implying that the vessels can transmit chemical signals to neurons. There are also hints that the dilation of blood vessels during hyperemia could stimulate neurons via pressure sensors on the surfaces of some neurons. If true, this would be analogous to how our sense of touch works through pressure on the fingertips. A functional role for glia is also increasingly supported by recent neuroscience research. Selective activation of glia using a technique called *optogenetic stimulation* can alter both spontaneous and stimulus-induced firing rates of nearby neurons. Glial activity can even influence behavior. In one example, Ko Matsui and his group at the National Institute for Physiological Sciences in Japan showed in mice that stimulating glia in a brain region called the *cerebellum* affected eye movements previously thought to be coordinated only by neurons in this structure.

A particularly extraordinary example of the influence of nonneuronal brain components comes from work of Maiken Nedergaard at the University of Rochester. Her laboratory transplanted human glial progenitors—embryonic cells that mature into glia—into the forebrains of developing mice. When the mice grew into adults, their brains were rich in human glial cells. The animals were then analyzed in a test of their ability to associate a short tone with a subsequent mild electric shock. During this kind of procedure, animals exposed to the tone-shock pairing begin to react to the tone as they would normally react to a shock alone (usually by freezing); the "smarter" an animal is, the more quickly it learns that the tone predicts an impending shock. In this case, the mice that carried human glial cells performed three times better in the task than reference mice that only received glial transplants from other mice. The hybrid animals also learned to run mazes more than twice as fast as the reference mice and made about 30 percent fewer errors in a memory recall task. It is simplistic to suppose that the mice performed better because of something the new glia did all on their own, but the experiments nevertheless show that these uncharismatic cells can influence behavior in nontrivial ways. With this comes the astonishing suggestion that the secret to human cognitive success may lie at least partly in our once neglected glia.

I n the narrow fluid-filled alleys that snake between cells of the brain thrives another form of brain activity that defies typical notions of computation. It is in these interstices that much of the chemical life of the brain takes place. To some, the idea of chemistry in the brain might evoke the psychedelic experiences induced by LSD and cannabis, but to neuroscientists, the term *brain chemistry* refers primarily to the study of neurotransmitters and related molecules called *neuromodulators*. Most communication between mammalian brain cells relies heavily on neurotransmitters secreted by neurons at synapses. Neurotransmitters are released when a presynaptic neuron spikes, and they then act rapidly on postsynaptic neurons, via specialized molecular "catcher's mitts" called *neurotransmitter receptors*, to induce changes in the postsynaptic cell's probability of spiking. In this neuronocentric view of the brain, neurotransmitters are primarily a means for propagating electrical signals from one neuron to the next. To the extent that neuroelectricity is indeed the lingua franca of the brain, this view seems justified.

But now imagine an alternative *chemocentric* view in which the neurotransmitters are the main players. In this view, electrical signaling in neurons enables the spread of chemical signals, rather than the other way around. From a chemocentric perspective, even the electrical signals themselves might be recast as chemical processes, because of the ions they rely on. This picture is upside-down by the standards of contemporary neuroscience, but it has something going for it. Perhaps most obviously, neurotransmitters and their associated receptors play functionally distinct roles far more diverse than neuroelectricity per se; by some counts there are over a hundred kinds of transmitter in the mammalian brain, each acting on one or more receptor type. An action potential means different things depending on which neurotransmitters it causes to be released and where those transmitters act. In parts of the central nervous system such as the retina, neurotransmitters can be released without spiking at all.

Neurotransmitter effects are also shaped by factors that are independent of neurons. Glia exert substantial influence because of their role in scavenging some neurotransmitters after they have been released. If the rate of neurotransmitter uptake by glia changes, the amount of neurotransmitter would be regulated in much the same way that the level of water in a bathtub is affected by opening or closing the drain. Glia also release chemical signaling molecules of their own, sometimes called *gliotransmitters*. Like neurotransmitters, gliotransmitters can induce calcium signals in both neurons and other glia. The functional effects of gliotransmitters in behavior and cognition are significant topics of current research.

The action of neurochemicals is also heavily influenced by a cell-independent process called *diffusion*, the passive spreading of molecules that results from their random movement through liquids. Diffusion is what causes the spontaneous dispersion of oil droplets over the surface of a puddle, or the aimless dancing of microscopic particles in milk, known as Brownian motion. Diffusion also influences the postsynaptic activity of neurotransmitters in important ways that are not yet fully appreciated but that represent a stark contrast to the orderly communication of information across circuit-like contacts between neurons. Some neurotransmitters and most neuromodulators are known in particular for their ability to diffuse out of synapses and act remotely on cells that do not form direct connections to the cells that released them. One such diffusing molecule is dopamine, the neurotransmitter we saw previously in the context of reward-related learning in monkeys. The significance of dopamine diffusion is highlighted by the action of narcotics like cocaine, amphetamine, and Ritalin. These drugs block brain molecules whose job it is to remove dopamine after it has been released at synapses. In doing so, the drugs increase dopamine's tendency to spread through the brain and influence multiple cells.

Neurotransmitter diffusion also underlies the phenomenon of *synaptic cross-talk*, another unconventional mode of brain signaling whereby molecules released at one synapse trespass into other synapses and affect their function. From the invaded synapse's point of view, this is like hearing a third person's voice murmuring on the phone while trying to have a one-on-one conversation with a friend. A number of studies have documented surprising levels of cross-talk among synapses that use the neurotransmitter glutamate, which is released by 90 percent of neurons in the brain and is mainly known for fast action within individual synapses. Such results are remarkable because they challenge the notion of the synapse as the fundamental unit of brain processing. Instead, both synaptic cross-talk and the more general effects of neurochemical diffusion in the brain represent aspects of what is sometimes called *volume transmission*, because they act through volumes of tissue rather than specific connections between pairs of neurons. Volume transmission arises from overlapping ripples of fluctuating neurotransmitter concentrations that seem more like raindrops falling on a pond than like the orderly flow of electricity through wires.

So from the neurotransmitter's point of view, neurons are specialized cells that help shape neurochemical concentrations in space and time, along with glia and processes of passive diffusion. Neurotransmitters in turn influence cells of the brain to generate more neurotransmitters, both locally and remotely. Whenever a sensory stimulus is perceived or a decision is made, a flood of swirling neurotransmitters emerges, mixing with the background of chemical ingredients whose pattern constantly wavers across the brain's extracellular space. Seen through this murky chemical stew, the electrical properties of neurons seem almost irrelevant—any sufficiently rapid mechanism for interconverting chemical signals would do. Indeed, in the nervous systems of some small animals, such as the nematode worm *Caenorhabditis elegans*, electrical signals are far weaker, and action potentials have not been documented.

The brain seen in this way is more like the ancients' vision, with not four humors but rather a hundred vital substances vying for influence in the brain's extracellular halls of power, not to mention the thousands of substances at work inside each cell as well. This chemical brain is a mundane but biologically grounded counterpoint to the shining technological brain of the computer age, or to the ethereal brains actuated by quantum physics and statistical mechanics. We can imagine the chemical brain instead as a descendent of the primordial soup of protobiological reagents that first gave rise to life in the Archean environment of the young planet Earth. We can also imagine the chemical brain as a close cousin of the chemical liver, the chemical kidneys, and the chemical pancreas—the offal we eat, all organs whose function revolves around the generation and processing of fluids. In this way, the brain loses some of its mystique.

I am one of those unhappy souls who discovered Douglas Hofstadter's cult classic *Gödel, Escher, Bach* (*GEB*) late in life. When my college roommate tried to amuse me with the bedazzling puzzles peppered throughout this book, I kept my nose boorishly buried in my physics and chemistry homework. I finally picked up *GEB* years past the wilting of my salad days, when I had neither the patience nor the youthful agility to give the puzzles the attention they deserved. Although I love Bach, enjoy Escher, and remain intrigued by Gödel, my mind had closed too far to revel in the book's rather mystical musings about consciousness. In one chapter, Hofstadter explains the structure of the nervous system

FIGURE 4. The faces and vase illusion.

as he saw it in the 1970s, a factual summary surprisingly consonant with the science of today and indicative in some ways of how slowly the field of neuroscience has been progressing. The description is also thoroughly redolent of scientific dualism. Wholeheartedly embracing the computer analogy, Hofstadter hypothesizes that "every aspect of thinking can be viewed as a high-level description of a system which, on a low level, is governed by simple, even formal, rules."

Another passage in *GEB*, however, strongly resonates with the point I have tried to make in this chapter; it deals with the relationship between figure and ground in drawings and other art forms. Hofstadter discusses cases where the background can be a subject in its own right, a phenomenon seen most famously in images of a vase versus two faces in profile (see Figure 4). In modern neuroscience, neurons and neuroelectricity have constituted the brain's figure, while many other components of brain function make up the ground. This gestalt has contributed prominently to the computational interpretation of the brain and to the persistence of the brain-body dualism. But just as visual perception can seamlessly flit from vase to faces and back again, so our view of brain function could as easily shift to emphasizing nonneuronal and nonelectrical features that make the brain appear more akin to other organs. Chemicals and electricity, active signaling and passive diffusion, neurons and glia are all parts of the brain's mechanisms. Raising some of these components above the others is like choosing which gears in a clock are the most important. Rotating each gear will turn the others, and removing any gear will break the clock. For this reason, attempts to reduce cognitive processing to the brain's electrical signaling, or to its wiring—the neural fibers over which electrical signals propagate—are at best simplistic and at worst mistaken.

Our embrace of the notion that brains function according to exceptional or idealized principles, largely alien to the rest of biology, is a consequence of the cerebral mystique. Our brains seem most foreign and mysterious to us when we imagine them as powerful computers, wondrous prostheses embedded in our skulls, rather than as the moist mixtures of flesh and fluids that throb there and also throughout the rest of our bodies. What better way to keep our souls abstract than to think of the organ of the soul as abstract, dry, and lacking in humor? We will see that this is but one of the ways in which idealization of the brain conflicts with a more naturalistic view in which the brain and mind are enmeshed in their biological and environmental context. In the next chapter, we will consider in particular how widespread emphasis on the extreme complexity of the brain contributes powerfully to the cerebral mystique and to the dualist distinction between brain and body.

three

IT'S COMPLICATED

FEW THINGS IN TODAY'S INTERNET-DOMINATED WORLD are as mysterious as the "it's complicated" relationship status on Facebook. Does "it's complicated" just mean that you're in an uncommitted sexual relationship? Does it mean you're in multiple relationships? Does it mean you're in the process of getting together or breaking up, but not sure which way it will go? Does it mean that you're cheating on someone? Whatever it means, posting "it's complicated" on a social networking website with over a billion users is an invitation to be asked—and presumably an excuse to give as equivocal and convoluted an answer as you can come up with. If you want to develop a mystique around yourself, "it's complicated" is definitely the relationship status for you!

"It's complicated," or words to that effect, is also a phrase you have probably heard about the human brain. Christof Koch, a leading neuroscientist who is chief scientific officer at the cutting-edge Allen Brain Institute, has called the brain "the most complex object in the known universe." His sentiment has been echoed by countless others. Neurobiologist and best-selling author David Eagleman quipped, "If our brains

51

were simple enough to be understood, we wouldn't be smart enough to understand them." "No computer comes close to its complexity," wrote journalist Alun Anderson in the pages of the *Economist*, "nor does the entire global communications network." "We won't be able to understand the brain. It is the most complex thing in the universe," commented Robin Murray, one of Britain's most prominent psychiatrists, on a BBC radio program in 2012. Even three hundred years ago, the famous French philosopher Voltaire supposedly remarked on the brain's intricacy with a cynical twist: "The human brain is a complex organ with the wonderful power of enabling man to find reasons for continuing to believe whatever it is that he wants to believe."

Could it be that the brain's complexity itself provides a shelter for some of our beliefs? If we don't want to question our conceptions of consciousness, individuality, and free will, the labyrinthine brain is a perfect hiding place. Assertions that the brain is unfathomably complicated might also create space for unfathomable influences on cognition, or they could justify nonscientific approaches to understanding the mind. "The more complicated a system is, the stronger it argues for having been intentionally designed," contends a writer from the Institute for Creation Research, which promotes biblically based pseudoscience. Lyricizing the brain's complexity might also serve to stoke interest in neurobiology or justify more funding for brain science. Or it might just be stating a fact. Regardless of the ends, emphasizing the brain's complexity serves to distance the organ from less mystifying aspects of the natural world, such as the biology of the rest of our bodies. In this chapter, I will discuss formidable intricacies of the brain but argue at the same time that the significance of this complexity is often overplayed. We should be able to grapple with the brain's complexity without abandoning a biologically grounded view of the mind.

Before the brain became well-known, the stars were where mysticism met complexity. The legendary Indian sage Vyasa described the heavenly bodies as a cosmic dolphin with Mars at its mouth, Saturn at its stern, the sun at its chest, and the moon as its mind. The belly of the dolphin was the "celestial Ganges," a streak of extraterrestrial luminance reflecting the earthly glistening of the holy river of Hinduism. To the ancient Greeks, the dolphin's belly was a milky continuum called Galaxias, passed into English via its Latin adaptation Via Lactea to give us the name Milky Way. Greek mythology had it that the milk was spilled when Hera pushed the suckling infant Heracles away from her bosom. For centuries, observers across the globe speculated that the Milky Way might be something more than a hazy smear in the nocturnal canopy, however. The eleventh-century Moorish scholar Avempace presciently hypothesized that the Milky Way is formed by light from many "fixed stars which almost touch one another." But it was not until the latter days of the European Renaissance that the galaxy's granularity could be directly observed. "By the aid of the telescope lately invented," wrote Galileo Galilei in 1610, "the very eyes of astronomers are conducted straight to a thorough survey of the substance of the Milky Way; and whoever enjoys this sight is compelled to confess that the Milky Way is nothing else but a mass of extremely small stars." Peering out into the night with revolutionary optical devices, Galileo resolved vast numbers of stars wherever he pointed his instruments. The enormity of his discovery has become a measuring stick by which to judge scale and complexity throughout the natural universe.

In contrast to a galaxy, the nervous system reveals its complexity when one shuffles the telescope's lenses and peers in at it through a microscope. One of the first people to observe the microscopic structure of the brain was the Bohemian anatomist Jan Purkyně (Johann Purkinje in the then-dominant Germanic spelling), who in 1838 reported his discovery of neurons in the cerebellum that now bear his name. In his original sketches, the Purkinje cells look like little overripe onions, each suggestively sprouting one or two shoots that taper mysteriously off into nothingness. Given the relatively crude optical systems of the time, Purkyně could only observe these cells because they are among the largest in the brain. Each of the bulbous forms he was looking at was the body or *soma* of a neuron—the fattest part of the cell—with a diameter of about one-thirtieth of a millimeter.

It was not until better lens systems and staining techniques came into use, later in the nineteenth century, that scientists could see where the onion shoots went, and the answer was phenomenal. From every Purkinje cell arises a bushy outgrowth of thousands of branching filaments called *dendrites*, each miniscule in diameter, but collectively spreading over many hundreds of times the volume of the cell soma. Each Purkinje neuron also gives rise to a single long root extending more than two centimeters through the brain tissue, called an *axon*. Similarly elaborate architectures are ubiquitous throughout the nervous system, as documented most famously in the detailed drawings of neuroanatomists Santiago Ramon y Cajal and Camillo Golgi. But even the impressive intricacy of single nerve cells becomes almost insignificant when one considers the sheer number of such cells in a human brain.

The case for complexity is indeed most often made with numbers. The love life of Mozart's operatic hero Don Giovanni is complicated because he has had 2,065 lovers, as his valet explains to a distraught paramour in the so-called Catalog Aria. (Imagine what Giovanni's Facebook page would have said about that.) The ATLAS subatomic particle detector, which provided some of the first direct evidence for the elusive Higgs boson, is complicated because it has about a hundred million readout channels, constructed over five years by a team of about three thousand physicists. The detector processes something on the order of a billion events per second, each of which generates around 1.6 megabytes of data. We now know that Galileo's Milky Way consists of somewhere around three hundred billion stars; the entire universe probably contains over two hundred billion times as many—leading to a total of about seventy billion trillion (seven with twenty-two zeros). Astronomer Carl Sagan's familiar catchphrase "billions and billions" hardly does justice to these numbers. How does the brain stack up against these benchmarks?

Quantifying the complexity of the brain with hard numbers actually takes a great deal of effort. Counting cells is the most obvious approach, but it is almost impossible to count cells throughout an entire brain using standard tissue analysis methods. To take on the task, the Brazilian neuroscientist Suzana Herculano-Houzel developed an industrial procedure. Her laboratory obtains brains of freshly deceased subjects and uses a combination of corrosive chemicals and mechanical mastication to reduce them to a slimy, viscous liquid. An important part of each brain cell survives this destruction, however—the *nucleus*, which holds the DNA of each cell and can also be used to identify whether its source was a neuron or a glial cell. Researchers can then count the density of nuclei in a known volume of brain-derived sludge, and thereby determine the number of cells that once made up the dissolved organ. Herculano-Houzel and her colleagues used the technique to count an average of about 171 billion cells per human brain, of which about half are neurons.

A much more laborious procedure can be used to estimate the number of synapses. Scientists carefully stain pieces of postmortem brain with a metallic chemical that sticks particularly well to synapses. They then cut the brain tissue into delicate slices, each less than a thousandth of a millimeter thick, and examine the slices with fifty-thousand-fold magnification under an electron microscope. By counting synapses this way in many representative slices, the researchers can extrapolate a value for the average number of synapses in brain regions from which the slices were cut. This type of process indicates that there may be as many as ten thousand synapses per neuron in the human cortex.

What do these cell and synapse numbers imply about the brain's capabilities? If we indulge in a simplistic computational analogy for a moment and imagine that every synapse is comparable to a computer bit—a switch with two positions for 0 and 1, depending on whether the synapse is inactive or active—the brain would have a storage capacity of about a hundred thousand gigabytes, roughly the scale required to store twenty thousand feature-length films at today's high-definition standard. (Think about putting all of Netflix in your head, and you would get a sense of the magnitude involved.) But the brain is not a disk drive; its vast reservoir of synapses is used primarily for data transmission between cells, a process that also changes the strength of each synapse. Many synapses are engaged and updated several times a second. You could not really store all of Netflix in your brain, but your trillions of synapses can support functions far more dynamic and varied than the kind of device that could.

Compounding brain complexity at the level of cells and synapses is a microscopic universe of sophisticated elements within each cell. Every cell carries the thirty-five thousand genes we humans are equipped with. Genes are turned on (expressed) in profiles that vary substantially across brain structures; in mice, gene expression patterns alone allow us to identify over fifty contiguous brain regions and subregions. Each brain cell also carries numerous organelles, the subcellular structures that do things like store genetic material and digest waste. Mitochondria, the "powerplants of the cell," are particularly abundant organelles in the brain, where they consume around 20 percent of the entire energy supply used by our bodies. At an even smaller scale, the brain contains countless numbers of bioactive molecules. Important classes of such molecules include the hundred or so neurotransmitters and neuromodulators discussed in the previous chapter, as well as large biological molecules, like proteins and DNA, that perform most specialized functions within each cell. Altogether, there are more molecules in the brain than stars in the universe—literally billions of billions of billions.

Many neuroscientists would say, however, that the complexity of the brain is most dramatically represented not by the number of its components but by the interactions among them. A bucket of water contains more molecules than the brain, but because each molecule in the bucket has the same monotonous H_2O formula, only a relatively small number of distinct types of interactions can take place. In contrast, biomolecules of the brain possess many distinct detailed structures that undergo selective, shape-dependent interactions with specific sets of other molecules. If every type of molecule in the brain were represented by a dot and every interaction were a line running between pairs of dots, the result would be a giant fur ball of overlapping lines that would require advanced computational analysis to interpret.

This molecular complexity within cells is present within all organs of the body, but interactions *between* cells add an additional layer of complexity that is peculiar to the brain. The slender axons and dendrites of neurons, as well as the tentacle-like cellular processes of astrocytes, allow brain cells to reach out and touch scores of distinct cells simultaneously. Individual neurons sometimes have hundreds of these little projections, which act as cables down which electrical impulses pass. Axons that carry information from one part of the brain to another may be several centimeters in length and make up the pale core of the cerebral cortex, called *white matter*. According to some estimates, the total fiber length in white matter exceeds a hundred thousand kilometers in normal adults, over twice the circumference of Earth and greater than the length of roadways in the entire US interstate highway system. For contrast, consider the liver, which contains as many cells as the brain but with far more restricted connectivity. Liver cells are compact and contact only about a dozen of their immediate neighbors in the tissue. They live in an era before roads and telephones, compared with brain cells that inhabit the internet age.

The task of mapping all the connections among cells in the brain would terrify even a scientific Hercules, but this is exactly the mission of a relatively new area of neuroscience called *connectomics*. Connectomics researchers have implemented on a massive scale the same types of procedures used to count synapses in electron microscopy. Instead of examining individual ultrathin brain slices, however, these scientists systematically examine every slice—including every cell and every synapse—in whole blocks of tissue. Because of the cost and difficulty of doing this, only tissue blocks much less than a cubic millimeter in size have so far been analyzed, but new information about cell-to-cell contacts is already being learned.

In one of the first published connectomics studies, Winfried Denk, Sebastian Seung, and their colleagues analyzed a small piece of a mouse retina; although the retina is not literally part of the brain, it is anatomically very similar to brain tissue and is also considered to be part of the central nervous system. Using a mixture of automatic data processing and twenty thousand hours of manual image analysis (fortunately divided among many people), they outlined 840 neurons within the retinal tissue block. Each neuron contacted an average of about 150 of the other cells, close to the number of online friends a typical Facebook user has. Think about what this means for the number of possible contacts among the 100 billion neurons in the human brain: If each of these neurons could contact 150 randomly chosen partners, then a single cell alone would have about $10^{1,389}$ (one followed by 1,389 zeros) possible configurations. This number dwarfs any quantity we've encountered in nature; even the number of atoms in the known universe is thought to be a paltry 10^{80} (one followed by eighty zeros). Although counting configurations like this is a very artificial way to think about brain structure, the result illustrates the astounding versatility that connectivity patterns can theoretically give rise to.

It is easy to embrace the cerebral mystique as we contemplate the brain's immense complexity in numerical terms. Overcome by its intricacy, we might justifiably view the brain as a riddle wrapped in a mystery inside an enigma. Regardless of whether it is a product of creation or evolution, how could we ever figure out how it works? If we are tempted to give up hope that the brain and all its wonderful properties will ever be understood, it may be because we think that the task of grappling with the billions of cells, trillions of connections, and octillions of molecules is simply too formidable for human ingenuity to conquer.

But before we despair, let's question to what extent the astronomical numbers of cells and connections in the human brain are actually necessary for explaining its function. If a drop is removed from a bucket of water, there is barely a difference—we could even describe the bucket's contents in physical terms that have nothing to do with individual droplets. Similarly, shouldn't we ask to what extent individual brain cells and their connections matter to the function of the brain?

It turns out that there are some surprising answers. One comes from looking at brain sizes. Normal adult brain sizes vary by 50 percent, from about 1 to 1.5 liters. Brain volume correlates only weakly with intelligence, however, reportedly accounting for only about 10 percent of the variability in IQ. Although some of the disparities in brain size can come from differences in cell density, size also correlates with variations in total brain cell number—at least in mice, for which data are available. This makes it likely that human brain sizes vary considerably in the number of cells and connections they contain, but that this variation does not strongly affect mental function. Alterations in brain cell number also occur during aging and disease, often without obvious consequences for cognition. Brain volume decreases annually by about 0.4 percent during normal aging and more than 2 percent per year for patients with Alzheimer's disease, even prior to diagnosis. This suggests that people can endure the death of billions of brain cells with no more than mild cognitive dysfunction. Clearly, not every cell in the brain is sacred.

A rare and remarkable congenital defect illustrates the expendability of brain cells even more emphatically. In 2014, a twenty-four-year-old woman entered a Chinese clinic complaining of nausea and dizziness. The woman had a history of balance problems, and had learned to walk and talk only by the relatively late age of seven. When the doctors performed a brain scan on her, they found that an entire region of her brain, the cerebellum, was missing. The cerebellum is involved in balance and coordination, and it happens also to be the most cell-dense part of the brain; it accounts for 10 percent of brain mass but contains 80 percent of all neurons—in this case all gone! Nevertheless, by her mid-twenties, this cerebellum-less woman was married with a child, and appeared to be living relatively normally, with only "mild mental impairment and medium motor deficits."

Similarly, radical brain disfiguration can result from intracranial surgery to relieve epilepsy. In extreme cases, doctors sometimes decide to remove an entire hemisphere of the cerebral cortex. This dangerous procedure puts the patient's life at risk and almost always results in paralysis of the opposite side of the body. But in other respects, the removal of huge chunks of the brain can be surprisingly well tolerated. A group of surgeons at Johns Hopkins Medical School performed fifty-eight hemispherectomy operations on children over a thirty-year period. "We were awed," they wrote later of their experiences, "by the apparent retention of memory after removal of half of the brain, either half, and by the retention of the child's personality and sense of humor." The results are all the more notable because, unlike the cerebellum, the brain regions removed by hemispherectomy are the most closely associated with human cognition, and the best developed in people versus other animals. Examples such as these reveal the extent to which redundancy reigns inside the head. Enormous parts of the brain can be missing, killed off, or removed without compromising essential aspects of personality and thought.

Even a human brain that lacks 80 percent of its neurons still retains vast numbers of components—after all, what's a few tens of billions of cells if there are a hundred billion to go around? Injured and malformed brains are like countries hit by war or famine. Their populations are depleted, but they still remain of the same *order of magnitude* as healthy brains, and they presumably retain many intact biological mechanisms. The intimidating complexity of what survives in an injured brain may still be required for perception and cognition. But could even more brain parts be taken away without sacrificing basic aspects of its function? There is currently no way to determine the minimal number of parts a working human brain can get by on, but we can draw some inferences from our relatives in the evolutionary family tree. If we consider other animals, we could ask whether cognitive abilities tend to require billions of neurons. Nature's answer seems to be an unqualified no.

In striking contrast to their reputation as birdbrains, our avian cousins provide some of the best evidence that a small cranium can support rather sophisticated behavior. Jesus famously belittled the birds by insisting that they "sow not, neither do they reap, nor gather into barns" (Matthew 6:26), but one example of avian mental powers seems to disprove his point explicitly. Some birds actually *do* perform activities that resemble the farmer harvesting and storing grain, thereby demonstrating their ability to strategize, plan, and remember. The distinguished naturalist Charles Abbott documented a remarkable instance of such food hoarding by crows near a beach in 1883. "I have witnessed several times the occurrence of crows breaking mussels by dropping them from considerable heights," he wrote. "The mussels so dropped were left undisturbed until the returning waters [of the tide] made fishing impracticable, when the birds hastened to feast on the results of their intelligent labor. Marvelous as it may seem, these crows recognized the nature of the tides, and . . . made as good use of it as possible."

More generally, members of the corvid family, which includes ravens, magpies, and the crows Abbott admired, display feats of intelligence otherwise almost unheard of for animals outside the primate order. Corvids have been shown to anticipate and prepare for the future, make and use tools, and recognize people on the street or themselves in a mirror. The prophetic raven of Edgar Allan Poe's narrative poem, the mischievous thieving magpie of Rossini's opera *La Gazza Ladra*, and the clever crow who figures out how to drink water from a deep pitcher in one of Aesop's best-known fables all pay homage to the astonishing skills of these birds. Parrots are also well-known for their perspicacity. Since ancient times, they have been prized for their ability to communicate vocally and obey simple commands. In a particularly notable instance, ethologist Irene Pepperberg carefully recorded the amazing intellectual feats of Alex, an African gray

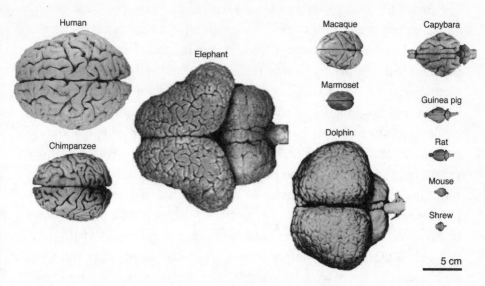

Figure 5. A comparison of brains from different species, shown to scale. Brain images come from the University of Wisconsin and Michigan State Comparative Mammalian Brain Collections (www.brainmuseum.org), which was funded by the US National Science Foundation and National Institutes of Health.

parrot she worked with for thirty years. Alex learned over a hundred English words and also knew how to count, categorize, and criticize. "You turkey!" he would call after people or things that displeased him.

The punch line to these anecdotes is that corvids and parrots are able to do such interesting things with brain sizes of only seven to ten milliliters, less than 1 percent of the cerebral volume of a person. These animals could not themselves have composed the poems of Poe or the arias of Rossini, but their abilities are ranked alongside those of chimpanzees and gorillas, our nearest evolutionary relatives, which have neuron counts about twenty times greater. Although we humans outrank all of these species in both intellectual capacity and number of brain cells, our brains in turn are outstripped by those of whales and elephants, which can have three to five times our brain mass but are generally considered to be significantly less intelligent (see Figure 5). This means that the absolute size of a brain or the number of its parts cannot hold the secret to understanding how brains mediate cognition and behavior.

The relative unimportance of brain size can also be seen by comparing members of species groups that vary radically in dimensions but relatively little in other respects. Consider the rodents, which range from the puny African pygmy mouse (about 8 grams) to the piglike Amazonian capybara (40–60 kilograms). These species both subsist in similar habitats and have highly social lifestyles, and they do not appear to differ dramatically in intelligence. Yet the capybara's brain weighs about 80 grams and contains about 1.6 billion neurons, while the pygmy mouse's weighs less than 0.3 grams and probably holds fewer than 60 million neurons. The rough correspondence of brain weight with body weight between these related species is not surprising—in fact, the *ratio* of brain to body weight is sometimes used to predict a species's intelligence. But this notion is also sharply at odds with the theory that the most sophisticated brains have the largest sizes and most cells. The brain/body ratios of the diminutive pygmy mouse and the giant capybara are roughly 1/20 and 1/500, respectively. According to this calculation, little David actually fares far better than Goliath, despite having only 4 percent as many neurons.

Studies across many species show that in general, smaller animals tend to have smaller brains but larger brain/body ratios. Perhaps this is why bigger animals are always outsmarted in cartoons. An ant's brain weighs one-seventh of its body, whereas our own human brains weigh one-fortieth of ours, predicting that ants might be about six times more intelligent than we are, if the brain/body ratio is what matters most. Researchers have gotten around the obvious problem with this kind of conclusion by observing that brains expand (or contract) at a different rate from bodies, in a manner that varies between branches of the evolutionary tree. This so-called *allometric scaling* of brains means, for instance, that if a certain type of species evolves over time to triple its weight, its brain size or number of neurons might only double. If further evolution toward larger size occurs, another tripling in body weight would correspond to another doubling in brain volume or neuron count. Because of allometric scaling principles, the large brain/body ratio of an ant probably cannot produce as much cognitive power as it would in a hypothetical human-sized insect. But the ant's tiny organ still supports an impressive behavioral repertoire, prompting no less than Charles Darwin to remark that "the brain of an ant is one of the most marvellous atoms of matter in the world, perhaps more so than the brain of man."

The scaling relationships between brain and body indicate that having billions and billions of brain cells is not necessarily beneficial in itself. Rather, having a big brain goes with having a big body, and natural selection may favor this combination for reasons having little to do with cleverness. Some biologists propose that comparative advantage in intelligence comes instead from having a brain that is *bigger than expected*, once scaling rules have been taken into account. By this type of measure, humans and other primates do well, with brain sizes and neuronal densities that are larger than those of other similarly sized mammals. But even in this regime, the generality of scaling principles supports the conclusion that brains with hugely different sizes can fall within the same IQ bracket.

So if cerebral volume and number of cells are not the keys to understanding brainpower, what is? In the last chapter, we saw an intriguing hint that growing human glial cells in mouse brains might make mice smarter. If there is something special about human glial cells, could there be other cell types that differ between species and help determine what the brains of each organism can do? In fact, many neuroscientists believe that brains are composed of a relatively manageable set of cell types, defined by the neurochemicals they use and the types of connections they make. Imagine the cell types as members of a construction team: excavators, bricklayers, plasterers, roofers, plumbers, and electricians. If the role of each type remains roughly the same in different parts of the brain, then the task of understanding brain function could be dramatically simplified—much as the construction of a city could be understood in large part by grasping how individual buildings are put together. Researchers are now trying to figure out exactly how many neuronal and glial cell types there are and what they do. Neurobiologists are also identifying and learning about characteristic structures formed by the different cell types. One such structure is called a *cortical column*. Columns are multicellular units about half a millimeter in diameter that cover surfaces of the brain like tiles of a mosaic.

The importance of building blocks like cortical columns and cell types suggests that major aspects of brain function might be figured out without reference to the vast numbers of such components. This approach has worked well for other organs. Human kidneys, for instance, contain more cells than the brain's cerebral cortex, but most are organized into millions of virtually identical structures called *nephrons*, which operate in parallel to one another to filter blood and eliminate waste. The pancreas also contains billions of cells, but pancreatic function can be analyzed in terms of a small set of well-defined cell types that produce each of the individual hormones the pancreas is known for. The ongoing experimental analysis of structural and functional subdivisions of the brain is a source for great optimism as we seek to understand how brains carry out their functions. Along with the evidence from small or disfigured brains, these possibilities for simplification pose a strong challenge to the glassy-eyed view that the complexity of the human brain, as described in simple numerical terms, places it outside the realm of nature or beyond the reach of science.

"What I cannot create, I do not understand." These words were found written on the blackboard of the Nobel Prize–winning physicist and nerd icon Richard Feynman at the time of his death in 1988. Some people have cited Feynman's epigraph as a goal we must achieve in order to claim victory in understanding the brain. Creating a brain could mean assembling it physically from cells in a laboratory or successfully simulating it on a computer. The billion-dollar European initiative called the Human Brain Project is now trying to simulate a brain by approximating the aggregate behavior of a hundred billion "virtual neurons" in computers. Related efforts in the United States aim to detect "every spike from every neuron" in a mammalian brain. Many neuroscientists are skeptical of such projects because they feel that the field has not yet progressed to an extent that justifies ambitions on this scale. In fact, computational biologists have yet to successfully simulate behavior of single biological molecules or cells, let alone organs, and experimentalists currently can barely record activity from even a few hundred cells in deep regions of the brain, let alone every cell. Given the state of today's science, mounting a project to simulate or monitor an entire brain at cellular resolution might be like trying to send astronauts to other galaxies before we've even been able to make it to Mars.

Big neuroscience endeavors that aim to understand the brain by modeling or measuring every cell reflect a preoccupation with the brain's complexity that has few precedents in other fields of study. If we really had comprehensive measurements of the human brain's structure and activity, we might well be able to shed light on its workings, but it seems just as likely that we might miss the forest (or at least large parts of it) for the trees. Imagine trying to analyze a great historical event like the French Revolution by following the anonymous movements of every single citizen in every house and street of the land. If we methodically plowed through all the data from 1789 to 1799, we would probably be able to read the changing social temperatures and detect flashpoints of tumult, but who can tell whether we would be able to identify key players—Danton and Robespierre, the Jacobins and Girondins, King Louis as he fled Paris in disguise—or explain their significance, as we became distracted by the antics of some of the other twenty-eight million people of France. The larger population might better be lumped together into the classes and estates (think cell types) that acted as collectives to bring about social change.

One of the few organisms for which we have nearly comprehensive neural data today is the lowly nematode worm, but we still lack the ability to "create" this animal's brain in any sense. Scientists today can measure the activity and connectivity of every cell in the worm's nervous system, but efforts to simulate its behavior are still rudimentary. Many researchers would probably agree that the most important insights into nematode neurobiology have come not from analysis of the comprehensive datasets but rather from targeted experiments in which specific behaviors such as crawling or egg laying were associated with a few distinct cells, genes, or signaling pathways. In other areas of biology, data about genetic makeup, which genes are turned on and off, and interactions among gene products (proteins) in individual cells have recently become available on a comprehensive scale. This so-called *omic*-scale data now enables scientists to study how huge numbers of molecules work together in processes like cell growth and communication. But even information like this tends to have the greatest impact when researchers can narrow in on a small number of factors that seem particularly interesting; they can then be probed in greater detail to test their importance more incisively.

Examples like these illustrate the fallacy of conflating data with *understanding*. Exhaustive information gathering does not necessarily lead to understanding, and understanding does not necessarily depend on all or even most of the data we could obtain and analyze. Think about understanding a car, for example. If you drive, chances are you've learned something about how cars work. A car with a standard internal combustion engine ignites gas in cylinders that expand, rotating a crankshaft, that subsequently transmits power to the wheels and makes the car move. Just because you understand how the car works at this fundamental level doesn't mean that you could repair or reassemble it from parts; for this we usually need mechanics. Conversely, if you saw the plan of a modern car, or even a movie of the car's mechanisms all in action, you probably couldn't pick out the function of most of the components, even if you could recognize a few key elements. Using the plans to simulate the car would be even more complicated—to do it well, you would probably need a host of information about unseen factors like friction, combustion efficiencies, and heat propagation that go far beyond the essential basics of how a car works.

In contrast to figuring out the workings of a car, achieving a holistic understanding of the brain is actually an ill-posed problem in the first place. A car, after all, has a single basic and self-contained function as a vehicle for transporting its passengers. The brain, on the other hand, is a multifaceted, multipurpose entity that can't operate apart from the organism it is part of. How the brain supports consciousness may be largely distinct from how it guides decisions, goes to sleep, or gets a seizure. Consider your own experiences. While you are having a casual conversation with a friend, you might also be looking out the window, watching trees swaying in the wind, recalling a line of poetry, and unwinding after a stressful day at work; but how you interact with your friend has little to do with how you perceive the trees or remember the verses, or how your mood changes as you decompress. We could try to explain each without the other, in fact. Rather different mechanisms lie behind the brain's various roles in domains such as communication, visual perception, and emotional regulation. It is likely that they can each be separately understood to a large extent, and indeed we have a great deal of rudimentary knowledge about these different processes already.

To ask that neuroscience studies account for all of the brain's functions at the level of individual cells, synapses, or molecules is to hold this organ to a special standard all its own. Doing so sets an almost unreachable goal for researchers, and one that is likely neither necessary nor sufficient for achieving meaningful understanding of the brain's many and disparate tasks. As we have seen, many brain parts may not even be required for generating its core capabilities. Although brains have intricate and enigmatic features, the fact that they are highly complex in numerical terms does not genuinely distinguish them from other products of nature or from other parts of the body. To cloak brains behind complexity is therefore to segregate them arbitrarily from the rest—this is the brain-body distinction in another form.

To experience human complexity firsthand, try Tokyo. With a sprawl embracing over thirty million residents, an economy surpassing all but a handful of countries, and a streetscape resembling a never-ending jumble of children's blocks, Tokyo remains the world's largest urban area. The city's rise from a small fishing village to a great metropolis, twice destroyed and twice remade into what we see today, is a remarkable testament to the sociotechnical success of our species in modern times.

To view our complexity in another way, visit the Sistine Chapel in Rome. For over five hundred years, this room has served as a private sanctuary for the world's most influential religious leaders. It is also one of the world's greatest artistic achievements—Renaissance luminaries labored for decades to execute the chapel's floor-to-ceiling frescos, culminating in Michelangelo's stunning renditions of the Creation and the Last Judgment, works without parallel in Western civilization. Like no other place, the Sistine Chapel expresses humanity's ability to transcend the bounds of animal nature.

To sample human complexity in a third way, just surf the web. The revolution of internet-age information technology makes us secret sharers of the lives of over a billion people, drawn from each nation and culture across the globe. Nearly every event, every book, every work of art, every creative thought or mad outpouring that ever left its mark upon the world, and many that have not, are there for us to download.

If the sophistication of culture is a measure of each species, then humanity so far outclasses the creatures of the wild that all comparisons are moot. It is tempting to suppose that the exceptionally complex achievements of humankind must stem from equally exceptional complexity in our human minds and brains. If we expected our thought organs to be as complex compared with animal brains as our culture is compared with theirs, we might even have legitimate reason to despair at the possibility of ever understanding how our brains work.

Could this feeling contribute to a mystique of the human brain's complexity? Could our cultural supremacy give rise to a sense of human *neuroexceptionalism*? A related attitude inspired efforts by nineteenth-century scientists such as Georges Vacher de Lapouge and Samuel George Morton to correlate brain dimensions with cultural advances and apparent intelligence among different human ethnic groups. As we learned in Chapter 1, results of the time purported to demonstrate the superiority of white people over others, primarily based on differences in brain size. This work has been widely debunked and is now viewed as a form of scientific racism. Revisiting the same ideas with respect to the differences between humans and animals is less controversial, but may be questionable for some of the same reasons.

Culture and brain can be dissociated on an evolutionary time scale. Approximately human brains have existed for much longer than our sophisticated society has. *Homo sapiens* and our close relatives in the genus *Homo* have been around for well over a million years. Brain morphology was fairly constant among these ancestral humans, though size has varied somewhat. Neanderthals, who originated over two hundred thousand years ago, actually had larger brains than ours, whereas historically more recent pygmy humans discovered on the Indonesian island of Flores had brains a third our size. The sole cultural relics we have from most of human evolutionary history are simple stone or bone tools. The earliest art on record is only about a hundred thousand years old, and urbanization and agriculture date back to the Neolithic Revolution, a mere ten thousand years ago. Before these times, our ancestors may have been little more than up-and-coming animals who knew how to communicate and use tools somewhat better than crows.

Culture and brain can be dissociated in the present as well. Sophisticated modern lifestyles are possible with or without human brains, and having a human brain certainly does not require one to interact with advanced technology. Even today, some human societies thrive with little dependence on the complex accomplishments of global civilization, although their members have the same biology as we have. The "uncontacted" tribes of New Guinea and South America, for instance, still maintain Stone Age practices and are almost completely isolated from more modernized societies. On the other hand, many nonhuman animals in our laboratories, zoos, and households live thoroughly enmeshed in twenty-first-century technology. Our domesticated friends with supposedly less complex brains benefit from modern medicine, eat processed food, admire and pose for pictures, and interact with a variety of electronic devices, just as we do. Of course our pets and laboratory animals were given access to technology by others; they are not innovating on their own—but then again, the same can be said for most humans.

The erroneous tendency to infer brain complexity from cultural complexity foreshadows themes we will take up in the following chapters, as we turn from considering myths about *what the brain is made of* to misconceptions about *how the brain interacts* with the body and environment. We will encounter brain-body distinctions as definite as those we have seen so far, and we will increasingly see how these ideas skew our attitudes about the human mind and soul.

four

SCANNING FOR GODOT

O NE OF THE GREATEST ADVANCES IN RECENT MEDICAL
technology has been the development of neuroimaging, the set of
procedures that lets doctors and scientists scan the living contents of the
skull without surgery. These procedures have had huge scientific impact,
garnered two Nobel Prizes, and probably done more than any other aspect of contemporary neuroscience to shape popular conceptions of the
brain. Judging from its prevalence and influence, neuroimaging seems to
have a mystique all to itself. Over ten thousand medical research articles
involving neuroimaging are now published each year. Brain scanning also
figures in fields as far-flung as economics and the law. You have almost
certainly seen brain images, perhaps spinning in 3-D to show you the
location of a tumor, perhaps sparkling with colors that show you how
the brain is affected by a task or treatment. If you have been to the hospital with any sort of neurological complaint, you have probably had an
X-ray computed tomography (CT) scan or a magnetic resonance imaging
(MRI) examination; the spinning brain or sparkling colors might have
been yours. Neuroimaging is how many people get to know their brains.

The most exciting form of neuroimaging is so-called *functional* brain scanning—the use of imaging to measure the brain in action, rather than merely its structure. This is most commonly performed with a functional MRI or *fMRI*, a technique I have spent much of my own career working with. In the 1990s, fMRI emerged as the most powerful method for mapping human brain activity, and it has since become a mainstay of neuroscience programs everywhere. To perform an fMRI experiment, researchers obtain a movie-like series of brain scans of a person lying in an imaging machine for a period of time. They then analyze the image series for time-varying changes that correlate with whatever the subject was doing or experiencing. These changes indicate how different parts of the brain are involved in the subject's behavior. Using fMRI, researchers have been able to find regions of the brain that process shapes, colors, smells, tastes, mistakes, actions, emotions, calculations, and much, much more. More edgy studies have sought to find brain areas whose activity depends on thinking like a lawyer or preferring Pepsi versus Coke. In the clinic, doctors and scientists apply fMRI and related methods to discern brain activity abnormalities associated with diseases like autism or schizophrenia. "The influence of fMRI-based research has changed our world," says Bruce Rosen, one of the inventors of the technique.

Befitting its importance in science and medicine, functional imaging is one of the most visible facets of brain research in the media and popular culture. Hundreds of newspaper articles related to fMRI are published each year, turning imaging into a flagship of modern neurobiology. Readers get hooked with catchy headlines like "This Is Your Brain on Politics" or "Watching New Love As It Sears the Brain." Particular attention goes to audacious suggestions that brain imaging can read minds, expose lies, and help marketers advertise their goods. Commenting on the public enthusiasm for such stories, psychiatrist Sally Satel and psychologist Scott Lilienfeld bemoan the fact that functional brain mapping has supplanted other valid ways to analyze mental and behavioral phenomena. At the same time, they admit that "it is easy to see why brain imaging would beguile almost anyone interested in pulling back the curtain on the mental lives of others."

By offering us face-to-face encounters with living human brains, one might also imagine that functional neuroimaging could provide an antidote to the cerebral mystique. What better way to grapple with the biological reality behind our minds than by seeing our own brains at work inside our heads, with the aid of techniques like fMRI? In a much discussed 2008 article, psychologists David McCabe and Alan Castel proposed that brain images fascinate laypeople precisely because they "provide a physical basis for abstract cognitive processes." But we will see in this chapter that the evidence in support of this view is sparse. Instead, brain imaging results are prone to contradictory interpretations that leave people free to choose among totally disparate concepts of the mind and brain. Even the most important and scientifically meaningful contribution of functional neuroimaging—the identification of brain areas specialized for distinct cognitive tasks—paradoxically reinforces the same kind of dualist perspectives we saw in the previous chapters. I will argue here that we must look well beyond the capabilities of today's brain imaging technology for a truer appreciation of the brain's place in human nature.

L et's start by testing your reaction to some real functional brain imaging data. A pair of mottled gray-on-black ellipsoids ogle you from the page (see Figure 6). The one on the right is dull and bluish, but the one on the left twinkles with hot sparks of orange and yellow. If you had been born yesterday, you might see these forms as nothing more than embellished Rorschach blots, ready to accept whatever significance you ascribe to them. But today, trained by countless similar images in the media, you probably know that they are brain images.

As it happens, the left and right ovals represent functional brain imaging data from two separate sets of subjects. The dull brain on the right side is labeled "obese group." The flashy one on the left is "nonobese." Its glint of color, absent on the other one, denotes activation of a brain region called the *prefrontal cortex*. Below them sits a caption: "When viewing images of food, obese participants had reduced activation in brain areas associated with self-control." The obvious implication of this vignette is that brain studies help explain how obese people react to food.

Can you believe these results? Do you find them intriguing, or perhaps surprising? Would you feel the same way if the caption was not

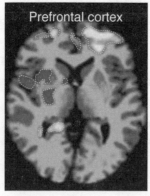

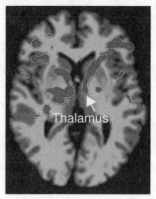

Nonobese group   Obese group

When viewing images of food, obese participants
had reduced activation in brain areas associated
with self-control.

FIGURE 6. A vignette including functional imaging data from Hook and Farah's study of the effects of neuroimaging on beliefs. Areas of activation are denoted by thin dotted lines; positive = light gray, negative = dark gray. Adapted with permission from C. J. Hook and M. J. Farah, "Look again: Effects of brain images and mind-brain dualism on lay evaluations of research," *Journal of Cognitive Neuroscience* 25 (2013): 1397–1405, © 2013 by the Massachusetts Institute of Technology.

paired with brain images? Do your beliefs about the mind or the soul influence your reactions? For instance, if you are religious, does your faith make you view the brain images or accompanying caption with lesser or greater skepticism?

Cognitive neuroscientists Cayce Hook and Martha Farah asked these questions in a large 2013 survey of how people respond to brain imaging. The researchers wanted in particular to test the McCabe and Castel hypothesis that brain imaging makes its splash by showing that mental processes have a physical realization. If the McCabe-Castel hypothesis were true, Hook and Farah reasoned, then people who believe in an incorporeal soul might tend to be more surprised and less accepting of neuroimaging results than others. Showing fMRI data to these dualists could be like introducing space aliens to someone who denies the possibility of extraterrestrial life. On the other hand, people who believe the mind is completely material should be unfazed by functional brain imaging. These people—often referred to as *physicalists*—probably already think the brain is where the magic happens and would not be surprised by more news of a relationship between brain and behavior.

Remarkably, Hook and Farah found that dualists and physicalists reacted similarly to the functional brain imaging data they saw. Vignettes that provided evidence for a physical manifestation of the mind in the brain did not appear to surprise or interest dualists more than physicalists, or vice versa. Moreover, the inclusion of actual brain images in the vignettes did not make much difference to how the participants responded, other than piquing both groups' interest a little more. "Across 988 participants," the researchers concluded, "we found little evidence of neuroimaging's . . . relation to self-professed dualistic beliefs." If brain imaging provides evidence of a biological basis for the mind, shouldn't it be otherwise?

It turns out that far from being shocked or disbelieving of neuroimaging data, some committed dualists actually embrace functional brain imaging as a tool for studying the disembodied mind they believe in. The Dalai Lama is one. Over the past decade, the spiritual leader of Tibetan Buddhism has worked with cognitive scientist Richard Davidson at the University of Wisconsin to organize a set of experiments in which the brains of Tibetan monks are scanned while they meditate. According to Buddhist dogma, meditation is one of eight steps along the spiritual path to attaining nirvana, a permanent escape from the endless cycle of birth, death, and reincarnation. But Davidson and his colleagues found plainly physical differences between the brain activation patterns detected from monks versus novices while they meditated. The results suggested that the monks' Buddhist exercises correlate with how their brains behave during meditation, and also seem quite compatible with the Dalai Lama's own dualist take. While physicalists might say that Davidson's brain imaging results show something about how brain activity underlies the act of meditation, and the mind more generally, the Dalai Lama simply turns this scenario on its head. He says he is interested "in the extent to which the mind itself, and specific subtle thoughts, may have an influence upon the brain."

Davidson's work with the meditating Tibetans is part of a research area sometimes called *neurotheology*, in which the neuroimaging and associated techniques are applied to analyze brain function during a variety of spiritual and religious activities. The very existence of this field depends on the compatibility between functional brain imaging and religion. Neurotheology laboratories have compared brain activity in religious believers and nonbelievers as they reason, moralize, or pray. These studies are sustained by cohorts of religious volunteers who presumably find that the experiments do little to challenge their traditional concepts of the soul. Andrew Newberg of the University of Pennsylvania has been a major promoter of neurotheology and has gained recognition for some of his functional neuroimaging work. In one study, Newberg's team recruited a group of devout Charismatic and Pentecostal Christians to engage in the euphoric practice of glossolalia (speaking in tongues) while being scanned. "I don't think faith has anything to be afraid of from science," explained pastor Gerry Stoltzfoos, who participated in the experiments. Stoltzfoos's attitude is very much like the Dalai Lama's: "Science validates faith," he insists.

A leading neurotheologist named Mario Beauregard goes beyond just studying spiritual phenomena in the brain. He has authored several books advocating nonmaterialist views of the mind, while at the same time using brain imaging to document neural correlates of mystical experiences. A 2007 *Scientific American* article characterizes Beauregard's work as "Searching for God in the Brain." Like the Dalai Lama, Beauregard fully accepts the brain's importance but views the brain as a servant to the disembodied mind. "A wealth of scientific studies," he argues, "indicate that our thoughts, beliefs, and emotions influence what is happening in our brains." In Beauregard's eyes, the fMRI machine is a tool for detecting spiritual influences on the matter of the brain, rather than for explaining the spirit itself in terms of matter.

Examples from the Hook and Farah survey to Beauregard's work show how easily functional brain imaging studies can be reconciled with supernatural beliefs about the mind or soul. It seems that if we use fMRI to pull back the curtain on our mental lives, we can see more or less what we want to. Although some people think the brain gives rise to the mind, and others think the mind controls the brain, nobody is surprised by the fact that the brain is involved. Even René Descartes, the man whose name became synonymous with dualism, posited that the spirit interacts with the body through a tiny brain structure called the *pineal gland*.

Neuroimaging has not provided the kind of information that would rule in or rule out such mind-brain interactions, and it therefore cannot serve as a basis for distinguishing among dualist or physicalist worldviews. To figure out why this is so, let us pull back the curtain on brain imaging itself, and take a closer look at the kinds of knowledge it actually produces.

Modern brain imaging was born in a drab hospital room in Wimbledon, near the site of the famous tennis tournaments. On October 1, 1971, a middleaged woman lay supine on an elevated stretcher, knees raised. Her head disappeared into a chunky square block, about a meter per side and twenty-five centimeters thick, held up edgewise by a heavy gantry. A cylindrical capsule attached to one edge of the square ran smoothly from corner to corner, like the lure at a dog track. After the capsule's run, the square rotated with a quick jerk about the patient's head. The capsule ran and the square rotated again and again, continuing rhythmically back and forth like cumbersome clockwork. After five minutes, the square had made half a revolution around the woman's head. In a room next door, stuffed with space-age electronic instruments, a picture flickered onto a computer screen: a white oval on a black background. The oval's dark, faintly textured core was bisected by a hazy light streak, but to one side a small black patch rudely disrupted the symmetry. The picture looked like a Miró painting, but it was actually the first clinical brain scan, obtained by Godfrey Hounsfield and his collaborators using their prototype X-ray CT scanner. The lady in the portrait had a brain tumor—the black patch in the oval. It was later successfully operated on, a coup brilliantly enabled by the newfangled neuroimaging know-how.

Although this experiment took place almost half a century ago, it debuted features of neuroimaging that persist to this day, such as the scanner configuration and the alliance of imaging hardware with computer processing. The basic idea behind CT is to measure X-ray transmission from every possible angle and position through a subject; in Hounsfield's setup this was accomplished with the dog lure and the rotating square, which contained an X-ray source and detector. A mathematical algorithm is then applied to reconstruct the image. CT images are static, however. In some cases, CT can help explain a cognitive problem by discovering the location of disruption, but it cannot indicate what the brain is doing during the scan.

The first brain scans to look dynamically at the brain's biological processes were methods sensitive to *radioactive tracers* (radiotracers for short). These substances are close analogs of natural biological or pharmacological molecules; when injected or ingested into the body, they go to the same places and do the same things as their nonradioactive counterparts. Radiotracers also emit gamma photons, which pass easily through biological tissue. Because this radiation can be detected non-invasively even with extraordinarily low tracer doses, there is minimal risk of side effects. Tracers called positron emitters yield two gamma photons simultaneously, enabling particularly sensitive and spatially accurate detection. Imaging these molecules in three dimensions became possible using the positron emission tomography (PET) scanner introduced in 1975 by Michel Ter-Pogossian, Michael Phelps, and their coworkers at Washington University in St. Louis.

PET imaging quickly formed the basis of several strategies for mapping aspects of brain function. In one approach, a positron-emitting version of the blood sugar glucose—the body's main energy source—is used to image cerebral metabolism. The radioactive agent ^{18}F-fluorodeoxyglucose (FDG) accumulates in proportion to the brain's glucose utilization. The buildup of FDG radioactivity can be monitored to see which areas of the brain are most active, at least according to their level of "fuel consumption." A second PET functional imaging method uses blood-borne radiotracers such as ^{15}O-water or ^{13}N-ammonia to measure changes in cerebral blood flow. Blood flow increases are evoked by neural activity and thereby result in greater delivery of the tracer to activated brain regions. Blood flow changes are harder to interpret in terms of neural mechanisms, but they take place faster than measurable changes in metabolic rate. In a further suite of applications, specific neurochemical processes are studied using radiotracers designed to interact with enzymes or receptors involved in those processes.

Many of the original PET brain imaging techniques are still in common use, and they have been joined by methods that use new tracers. A recent breakthrough, for instance, has been the development by the University of Pittsburgh's William Klunk and others of PET tracers that reveal the pathology of Alzheimer's disease. PET has proved limited for many studies of brain activity, however. One problem is that PET scans offer relatively coarse-grained spatial detail, or *resolution*. Pixel sizes of several millimeters are typical, meaning that every point in a PET scan corresponds to tens of thousands of cells and sometimes even more than one brain region. More importantly, PET scans are ponderously slow compared with brain processes like perceiving and thinking. Even the fastest functional PET experiments require scan times of about a minute—almost a thousand times longer than it takes to recognize a person's face, and about five times longer than it took world champion Magnus Carlsen to defeat Bill Gates in an entire game of speed chess.

Several of the weaknesses of PET were addressed by a radically different imaging technology developed by Paul Lauterbur at the State University of New York in 1973. Lauterbur was a chemist who specialized in an analytical technique called *nuclear magnetic resonance* (NMR) spectroscopy. NMR is an effect whereby the nuclei of certain atoms—most commonly the hydrogen atoms in water—absorb radio waves at specific transmission frequencies when placed in a strong magnetic field. Lauterbur discovered a way to use NMR to reveal the spatial positions of absorbing nuclei. Because biological tissues are largely transparent to radio waves (and undamaged by them), this new NMR-based imaging was perfect for visualizing living soft tissue in three dimensions. As NMR imaging gained steam in the medical community, the threatening N for nuclear was lost, and the method became best known as MRI. MRI was quickly recognized for its superior rendering of anatomical detail in soft tissues like the brain.

In the early 1990s, scientists discovered ways to perform functional brain imaging using MRI. In the first published fMRI study, Jack Belliveau, Bruce Rosen, and their collaborators at the Massachusetts General Hospital in Boston emulated earlier PET experiments by injecting an MRI contrast agent into the bloodstream of volunteer subjects while they were being scanned. The scientists were then able to map brain activity by seeing where the contrast agent accumulated during visual stimulation. At about the same time, another group of researchers led by Seiji Ogawa at Bell Labs showed that blood itself can act as an intrinsic contrast agent for fMRI. Because of the weak magnetism of both oxygen and iron in blood, small changes in blood flow and oxygenation can be detected without the need for injections. Such effects occur within seconds of elevated brain activity and are the basis of most current functional imaging experiments.

Not surprisingly, limitations of fMRI arise from its dependence on blood. The spatial resolution of fMRI is fundamentally limited by the spacing between blood vessels in the brain. This is somewhere around a tenth of a millimeter, much larger than brain cell sizes. Most fMRI signals are likely to arise from contributions involving many different types of neurons and glia, as well as possible changes in blood circulation unrelated to local brain activity. The hundreds of chemical messengers we considered in Chapter 2, as well as the synapses and connectivity of Chapter 3, are all lost in translation. Berkeley neuroimager Jack Gallant comments that "fMRI is like measuring the total electricity usage in your office at specific times to figure out what's going on at everyone's desk." Researchers also bemoan the sluggishness of fMRI compared with neural activity. Imagine watching a movie that has been blurred in time over several seconds per frame. Our favorite action heroes—Rocky and Ivan, Rosa and James, Obi-wan and Vader—would all be reduced to uninterpretable streaks of color. This is what blood flow effects detected by fMRI are like. For this reason, scientists sometimes complement fMRI with measurements from faster recording techniques called electroencephalography (EEG) and magnetoencephalography (MEG). But although MEG and EEG respond quickly to the brain's electromagnetic activity, they cannot localize the activity nearly as precisely or reliably as fMRI.

The signals detected by fMRI and other functional imaging approaches are also tiny—brain activity typically elicits ripples of at most a few percent in image intensity. These subtle changes play against a background of fluctuations resulting from the subject's movements, instability of the scanning equipment, and physiological processes unrelated to the study. Scientists must try hard to separate out image changes that are genuinely related to whatever stimulus or phenomenon they are trying to learn about. This usually involves expansive computational analysis of dozens of repeated trials, multiple subjects, and varied experimental conditions. The results of such calculations are usually depicted as colored blobs of inferred brain activity, superimposed on black and white anatomical images as in Figure 6. Such pictures provide the best information we can currently obtain about human brain function, but they do not really show what the brain is doing at any point in time, and they rarely come from anybody's brain in particular. Instead, functional brain maps are highly processed, statistical aggregations of image data that are sometimes as distant from underlying biological processes as bologna is from a pig.

The sophisticated computational tricks used in analyzing neuroimaging data come with equally sophisticated pitfalls. A University of California, Santa Barbara, postdoc named Craig Bennett shockingly illustrated this point by using apparently innocent fMRI methods to reveal brain activation in a dead salmon. Bennett and his collaborators scanned the late fish while it "saw" a series of photographs. A few pixels in the brain showed what looked like responses to the photos based on statistical testing, resulting in a very fishy version of a typical fMRI activation map. In reality, the apparent brain responses resulted from random fluctuations in the images, which were ineffectively screened out by commonplace analysis methods. Bennett had a difficult time publishing his satirical study but was aptly awarded an Ig Nobel prize, given to "honor achievements that first make people laugh, and then make them think." A second damning study, led by an MIT student named Ed Vul, revealed that high-profile brain imaging papers were frequently reporting statistically impossible results, the equivalent of claiming odds of better than 50-50 for a coin toss. These mistakes led authors of the offending studies to report correlations between brain areas and complex stimuli that were literally too good to be true. Although the types of errors that Bennett, Vul, and their colleagues exposed are neither inherent nor specific to brain imaging research, imaging studies are particularly prone to such problems because of the small signals and large datasets involved.

The indirectness and coarse resolution of current brain imaging techniques clearly provide wide latitude for disparate influences and interpretations. Examples like the salmon study also show us how easy it is to bias results simply by carelessly designing or analyzing experiments. We have seen that different people can look at imaging data and see the mind at work on the brain or the brain busy carrying out functions of the mind—brain images make as much sense to dualists as to physicalists.

After learning about what functional neuroimaging really tells us, we can begin to understand this contradiction. Today's brain activity maps are so indistinct that we can imagine almost anything going on behind the scenes.

The brain is like a Swiss Army knife, argues my MIT colleague Nancy Kanwisher, a pioneer in applying fMRI to problems in cognitive science. Despite limitations of the technique, imaging studies including hers have delineated a once unanticipated array of distinct brain regions that respond during specialized tasks, from recognizing faces to thinking about thinking (see Figure 7). Each of these brain areas seems to be specialized for its task, like different tools on the knife. Almost half of published neuroimaging studies are localization studies, and many of the rest deal with further characterization of defined brain regions. Localization results are the most obvious take-home lessons from functional neuroimaging research. When carefully performed and interpreted, they inform us about how the brain and mind are constructed, but when regarded superficially, the localization of cognitive functions can be a distraction from efforts to understand how our brains and minds really work.

Hard evidence for the specialization of brain regions has been around for a long time. Before PET and fMRI, data came largely from small numbers of neurological patients in whom specific cognitive or behavioral defects could be traced to focal brain injuries. Perhaps the best-known example was the case of a patient named Louis Leborgne, who was studied by the French physician Paul Broca in 1861. Leborgne had suffered from epilepsy since childhood, and as an adult lost his ability to speak; he was hospitalized and could utter only one syllable, "tan." Despite his speech loss, Leborgne's comprehension and general cognitive capabilities were unharmed—a set of conditions now known as Broca's aphasia. In a postmortem brain examination, Broca found that Leborgne had a lesion in the left frontal lobe of his cerebral cortex, and he went on to discover injuries at the same location in patients with similar language defects. Discovery of the correspondence between language production and the brain region called Broca's area represented a striking validation of Franz Gall's theory of functional localization (see Chapter 1). The

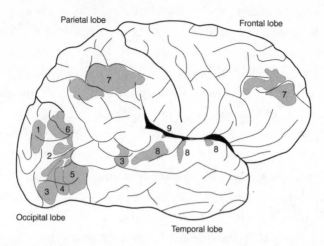

FIGURE 7. Human cerebral cortex showing lobes and areas that according to neuroimaging studies respond specifically to (1) places, (2) body parts, (3) faces, (4) faces and motion, (5) motion alone, (6) thinking about other people's thinking, (7) difficult cognitive tasks, (8) speech sounds, and (9) sounds with pitch.

underlying idea behind phrenology was at least partially correct, even if Gall's maps of specific regions and corresponding skull features were entirely wrongheaded.

Neuroimaging makes the same point but sharpens it. Using today's techniques, observations are uncoupled from the rare conjunction of serendipity and personal misfortune that led to breakthroughs like Broca's. Willing volunteers in generous numbers can be scanned during many different stimuli or tasks in single or multiple sittings. Researchers can see the results almost immediately after each experiment, rather than waiting for their subjects to die and undergo autopsy. The brains of healthy neuroimaging subjects are not distorted by damage or disease, as lesioned cerebra can be. Brain imaging results therefore usually reflect normal physiology. Most importantly, imaging, unlike injury, probes the entire brain at once. PET or fMRI can reveal if multiple structures are engaged in experimental paradigms and can characterize the extent and magnitude of responses in each region. The roles of structures involved in language, for instance, can all be examined in a single experiment: Broca's area, important for articulation; Wernicke's area, important for comprehension; auditory and motor cortices, important for generic aspects of hearing and movement; and numerous functionally important subdivisions of each of these regions are all involved and can be seen working in parallel during a speech task.

Discovery of specialized brain regions has indubitable biological significance. Just as forces at work throughout the earth's geological evolution gave rise to the mountain ranges, oceans, and rivers we see today, so we imagine that the factors that shaped human evolution patterned our brains in the ways we can now detect. The presence of brain areas or groups of areas whose activities correlate strongly with mental functions like speech and socializing suggests in particular that these functions are distinct adaptations with dedicated neural hardware, an interpretation now shared by many neuroscientists. "What's important . . . is not the particular locations of [the corresponding] brain regions," Kanwisher explains, "but the simple fact that we have selective, specific components of mind and brain in the first place."

But the importance neuroimaging studies place on associating mental faculties with physical spots in the brain has also led many to disparage it as a reincarnation of phrenological pseudoscience. "Critics feel that fMRI overlooks the networked or distributed nature of the brain's workings, emphasizing localized activity when it is the communication among regions that is most critical to mental function," writes David Dobbs in a *Scientific American* piece entitled "Fact or Phrenology?" Psychologist Russell Poldrack went so far as to compile a list of published fMRI studies that implicitly lend support to phrenological mental categories, illustrating the surprising compatibility between modern science and outmoded ideas. For each example, Poldrack related old phrenological classifications to thematically similar fMRI experiments, and then to the specific brain regions the imaging studies had identified. "One can be almost certain that Gall and his contemporaries would have taken these neuroimaging results as evidence for the biological reality of his proposed faculties," Poldrack comments. The way neuroimaging results are reported often strengthens Poldrack's point. Titles such as "Neural Correlates of Giving Support to a Loved One," "The Neural Substrate of Human Empathy," or "Neural Correlates of Superior Intelligence" give the impression that complex character traits can be boiled down to blotches of territory in the brain. It is easy to imagine some of the neural correlates and substrates—each corresponding to one or more localized fMRI activations—nestling alongside phrenological regions for attributes like "amativeness" and "acquisitiveness" on one of Lorenzo Fowler's ceramic heads.

Careless interpretation of brain localization results invites further criticism by implying that regional activity, like labeled bumps on the head, actually *stands for* particular cognitive processes. Advertising expert Martin Lindstrom claimed in a 2011 *New York Times* op-ed that people literally love their Apple iPhones because photographs of iPhones activate a brain region called the *insular cortex*, which is also among brain regions that respond when people view pictures of their romantic partners. Here Lindstrom treats insular cortex fMRI signals as denoting love, even though the truth is that this region responds during both negative and positive emotions. In his book *Imagine*, author Jonah Lehrer summarizes an experiment that found an association between problem solving and a brain area called the *anterior superior temporal gyrus* (aSTG). He writes that "the aSTG is able to discover" the answers to word puzzles—the aSTG itself becomes a problem solver. Even the Nobel Prize–winning biologist Francis Crick appears to slip into the same "brain area equals cognitive function" fallacy when he references lesion studies to argue that "Free Will is located in or near the anterior cingulate sulcus," a small fold near the midline of the brain.

This type of thinking is mistaken on both technical and theoretical grounds. The technical critique takes off from the limitations of brain imaging itself. Each blotch of activation represents thousands if not millions of cells, synapses, and neurochemicals, all contributing together to brain function like myriad voices engaged in a chaotic debate. Neuroimagers currently have no good way to parse or distinguish among the voices, so instead they tend to do the easiest thing: they listen to the loudest opinions! The loudest opinions could arise from unanimous stimulus responses of cells in a brain region, but more likely the viewpoints represent a majority, a plurality, or even just a vocal minority of cells overshadowing a silent majority. As far as fMRI and PET are concerned, the loudest voices are also those that produce the biggest blood flow changes, not necessarily the voices most important for actual brain function. A further complication is that brain activation maps are almost always determined by comparing imaging responses to a test condition with responses under one or more reference conditions, so that the brain regions identified are actually those that display greater activity under the test condition, rather than regions that are turned on *only* by the test. The upshot is that the fact that a particular brain region "lights up" most during a particular mental task does not mean that the region as a whole specializes in that task to the exclusion of other functions.

On the flip side, the regions that do light up in imaging experiments rarely include all of the brain areas involved in any given cognitive process. A key reason for this is the "tip of the iceberg" problem. Icebergs are known for being much larger than the parts you can see; 90 percent of each floating mass hides menacingly under the ocean's surface. In functional imaging analysis, the analog of the iceberg is a map of fMRI signal changes correlated with whatever task or stimulus was used in the experiment. The map is directly computed from the raw images. But although the map in principle covers the whole brain, the only parts you see as activation are areas where the reliability or magnitude of the imaging signals exceeds a cutoff value set by the experimenter. If the cutoff is set too low, too many peaks become visible, and the likelihood is high that some are caused by random nonneural fluctuations, like false activation in a dead salmon's brain. For typical conservative cutoff values, however, some of what remains hidden beneath the cutoff—the submerged part of the iceberg—is still due to task-specific brain function. This genuine brain activation is lost in the data analysis and usually never discussed. Because of this problem, most functional imaging studies systematically overstate the degree to which brain responses are localized to a few small areas.

James Haxby of Dartmouth College has argued that the whole iceberg, including imaging signals that are usually ignored, should be considered when interpreting brain scanning experiments. In an influential 2001 paper, Haxby and his colleagues strayed from the standard practice of focusing only on brain areas with the largest responses to experimental stimuli. By doing so, they were able to observe that neural responses reach across "a wide expanse of cortex in which both large- and small-amplitude responses carry information" about visual stimuli. This type of approach favors a picture of brain activity in which mental processes are distributed over much of the brain, rather than being compartmentalized in specific structures.

At a fundamental level, all neuroscientists know that a picture like this must be accurate. Even if a particular patch of brain displays highly specialized activity patterns, the activity has to come from somewhere. If a brain region is activated by faces, for instance, face stimuli have to percolate from the retina through multiple levels of the brain's visual system to generate signals in the brain area that responds most prominently to faces. If there were nothing face-like about the responses in other parts of the brain, then the face area would have no way to distinguish a face stimulus from any other. It could even be the case that *negative* fMRI signals outside the main face area—likely indicating reduced neural activity—help distinguish between faces and other stimuli in the brain. Such negative responses would be a bit like Arthur Conan Doyle's famous guard dog who didn't bark at an intruder in the night, revealing to Sherlock Holmes that the dog knew the intruder. But the variegated patterns of brain activity that might display such characteristics are necessarily overlooked by imaging analyses that focus only on peaks of brain activation and maximal responses.

This gets at the more theoretical problem with analyses that emphasize the location of cognitive processes: they tend to black-box the question of how the processes are actually performed. "Even if we could associate precisely defined cognitive functions in particular areas of the brain . . . it would tell us very little if anything about how the brain computes, represents, encodes, or instantiates psychological processes," writes psychologist William Uttal in his 2003 book *The New Phrenology*. In a similar vein, philosopher Daniel Dennett derides the idea that specific brain regions could account for one particular cognitive process: the phenomenon of human consciousness. Localizing the process merely recasts the problem of deciphering how the brain functions to deciphering how the particular brain region in question functions. Dennett sarcastically analogizes this approach to a theatrical performance in which the brain's embodiment of consciousness gets to "watch" all the mind's happenings and be conscious of them, a scenario reminiscent again of Descartes's mind-body dualism, even if it all takes place in the brain. Dennett regards this *Cartesian theater* as absurd because the brain region or regions that implement consciousness are as inscrutable as consciousness itself was to begin with, and the boundary between parts of the brain that are conscious and parts that aren't can be nothing but arbitrary.

When we emphasize the localization of other cognitive functions, we are setting the stage for analogously absurd theaters, in which the brain regions that perceive colors, sentences, regions of space, and so on become unnaturally empowered and separated from the parts that don't. Although no neuroscientist would endorse this burlesque view, it is exactly this impression that we get from simplistic portrayals of functional neuro-imaging results. To many researchers, the main value of localization stud-ies is not that they identify regions of the brain with apparently discrete functions, but that they provide clues as to how to begin more incisive experimental investigations, often using invasive experimental methods that directly probe cellular activity in animals.

Dennett's Cartesian theater analogy and its extension to other cog-nitive processes evokes the real-life theater of modernist playwrights like Samuel Beckett. In Beckett's absurdist masterpiece *Waiting for Godot*, the two vagabonds Vladimir and Estragon loiter by the side of the same road each day, hoping that they will eventually encounter the eponymous Godot, who never arrives. "You're sure it was here?" asks Estragon, won-dering if they are even looking in the right place for their man. Befitting a play that is often taken as a commentary about the meaninglessness of existence, Beckett and his critics never agreed on a single interpretation of the drama or characters. Like a colored blob of brain activation in an fMRI map, Godot in particular remains a mystery; we never learn who he is, what he represents, or whether he even exists. When we try to lo-calize cognitive functions using brain imaging techniques, are we perhaps scanning for Godot? Are we searching for an enigma sometimes defined more by our expectations than by reality, with little promise of enlight-enment at the end of our wait?

Contemporary neuroimaging strengthens the mystique of the brain with a combination of scientific luster, media hype, simple but sometimes simplistic findings, and compatibility with a wide variety of belief systems. With techniques like fMRI, we can learn interesting facts

about brain activity without any pressure to revise our deeply held attitudes. Those who hope that cognition can be demystified using today's
human brain imaging technology are out of luck, however. Even when
allied with the most sophisticated analysis methods, functional neuroimaging simply does not have the resolution or specificity to figure out
what brain activation patterns really mean, how they are established, or
how they are connected to the rest of the brain. "Claims that computational methods and non-invasive neuroimaging . . . should be sufficient
to understand brain function and disorders are . . . naive and utterly in-
correct," writes neuroscientist and fMRI expert Nikos Logothetis.

Functional neuroimaging results today are somewhat like cartography
before the days of authoritative atlases, settled borders, and satellite imagery. Like fMRI-based brain maps, ancient physical maps are often wonky,
inaccurate depictions, limited by the technology available at the time they
were made. Early cartographers found space for monsters as well as for
the landmasses we know today, while modern brain interpreters find areas
for free will as well as face perception. Some apparently specialized brain
regions will stand the tests of further exploration; for instance, the existence of dedicated areas involved in face recognition has been backed
up by electrode recordings, lesions, and stimulation studies in people and
monkeys. But some regions, and even the cognitive concepts that define
them, may prove as ephemeral as the lost lands of Thule and Atlantis.
Whether validated or not, the association of mental functions with cir-
cumscribed places in the brain by its very nature will continue to foster
a *neurosegregation* that keeps the biological bases of our mental processes
bounded, separated from each other and the rest of the world. If we want
to explain and understand at a deep level how mental functions work, we
must therefore look beyond today's human neuroimaging techniques.

Imagine a brain imaging method in which no cell's activity, no pathway or connection, and no swirl of neurochemicals go undetected. With
each touch, sound, or glimmer of light, a cascade of brain events takes
place, each one open to the all-seeing eyes of our "total neuroimaging"
technique. This is not a distant fantasy—it is at least approximate reality
in neuroscience laboratories that study small transparent organisms. By
combining cutting-edge optical microscopy techniques with fluorescent
biochemical indicators of neural activity, researchers like Misha Ahrens

of the Howard Hughes Medical Institute have already been able to record signals in almost every neuron of a baby zebrafish brain simultaneously. In experiments like his, there is vastly less room for uncertainty
about cause-effect relationships and the organization of neural activity
than there is in human brain imaging. Some scientists suggest that it may
one day be possible to adapt techniques analogous to Ahrens's for use
in humans. Some of my own laboratory's research aims, for instance, to
create biochemical neural activity indicators detectable by fMRI; these
could be a step toward total neuroimaging by allowing chemical and
cellular-level signals to be detected noninvasively.

Total neuroimaging is the kind of technology that in the future could
dramatically accelerate our ability to learn how the brain functions as an
integrated, multifunctional organ. Although we are a long way off from
having methods like this for studying people, progress is being made.
Even this would not enable us to understand how mental processes really work, however. Researchers have been using high-resolution whole-
brain optical imaging in worms and zebrafish for some time now, but as
we saw in the last chapter, supposedly comprehensive information from
simple nervous systems still falls short of explaining behavior. One of
the reasons for this is that the brain and nervous system do not perform
cognition on their own. Just as the brain regions discussed in this chapter
cannot be viewed as acting in isolation, so the brain as a whole cannot
be considered in isolation either; it must be viewed in the context of the
body and the environment. The next chapters will further explore the
continuum that unites the brain with its surroundings.

five

THINKING, OUTSIDE THE BOX

I N THE PREVIOUS CHAPTERS, WE SAW HOW THE BRAIN IS portrayed as the body's outlier. Through the lens of popular neuroscience (and even some not-so-popular neuroscience), brains become abstract and hypercomplex entities—mystical machines, rather than down-to-earth organs composed of flesh and blood. Von Neumann's brain-as-computer metaphor, exaggeration of the brain's complexity by journalists as well as scientists, and the tendency to black-box cognitive processes in neuroimaging studies all place the brain outside the range of normal biological phenomena. These trends exemplify the artificial brain-body distinctions I referred to as scientific dualism: conceptions of the physical brain that help preserve traditional attitudes about human nature, consciousness, and will, but that go against a more biologically realistic picture.

The dualist perspectives we have deconstructed so far relate to the makeup of the brain, how it is organized, and what makes it tick. But to many people, the brain seems exceptional not only in its makeup but also in its relationship to the world around it. "The brain is the control center of the body" is a statement we have all encountered. The implication is that the brain is like the CEO of a company or the captain of a ship. It is in charge. *Your Brain Is God,* declared Timothy Leary, prophet of the psychedelic neuroscience of the sixties, carrying the cerebrocratic viewpoint to an exuberant extreme. Other writers have argued for the centrality of the brain in more sober—but no less certain—terms. "All mental functions, from the most trivial reflex to the most sublime creative experience, come from the brain," says neurobiologist and Nobel laureate Eric Kandel, restating the ancient philosopher Hippocrates's assertion that cognitive functions arise "from nothing else but the brain."

Going a step further, Francis Crick posed what he termed the astonishing hypothesis that "'You' . . . are nothing more than the behavior of a vast assembly of nerve cells and their associated molecules." Here Crick actually *equates* the brain with the person it supposedly controls, as Shakespeare in his plays sometimes equates dukes and kings with their realms. The rest of the body seems virtually dispensable when compared with the hegemonic brain. Personification of the brain also takes place whenever we say things like "My brain is asleep" or "My brain can't take this anymore." Parts of the brain—regions or even individual cells—also become personified. Reporting on a study of neural responses in the human brain, a *Wall Street Journal* article describes "a neuron roused only by Ronald Reagan, another cell smitten by the actress Halle Berry, and a third devoted solely to Mother Teresa." Literary license is being used here, but the tendency to think of brain cells as doing what people do is also unmistakable.

Anthropomorphic descriptions of the brain and its components have become ubiquitous, but some philosophers believe they are also deeply erroneous. Through the mouth of his fictionalized prophet, Zarathustra, the philosopher Friedrich Nietzsche teaches: "Behind your thoughts and feelings, my brother, there is a mighty lord, an unknown sage—it is called self; it dwells in your body, it is your body." Nietzsche was rebelling against the mind-body distinctions of his intellectual forebears. But just as he rules out the possibility of a self that stands apart from the body, he resists the idea that the self could be contained in a particular component of the body. A less poetic but more precise expression of the same notion comes from Ludwig Wittgenstein, an icon of mid-twentieth-century scholarship. Wittgenstein writes in his *Philosophical Investigations* that "only of a human being and what resembles (behaves like) a living human being can one say: it has sensations; it sees, is blind; it hears, is deaf; is conscious or unconscious."

Speaking of brains or parts of brains as thinking, perceiving, or acting like living human beings violates Wittgenstein's dictum, argue philosopher Peter Hacker and neuroscientist Maxwell Bennett in their 2003 book, *Philosophical Foundations of Neuroscience*. To them, using psychological terms to describe what brains do is wrong because brains do *not* closely resemble complete people; language that personifies the brain instead represents a "mutant form" of the mind-body distinction left over from before explanations of the mind came to depend on neuroscience. "By speaking about the brain's thinking and reasoning, about one hemisphere's knowing something and not informing the other, about the brain's making decisions without the person's knowing, about rotating mental images in mental space, and so forth," write Bennett and Hacker, "neuroscientists are fostering a form of mystification and cultivating a neuro-mythology" that fails to advance public understanding or arrive at meaningful answers to questions about how the brain and mind work.

Such vehement rejection of "psychological brain talk" gets a mixed response from others in the field. Tufts University's Daniel Dennett is ready to accept some personification of the brain as both appropriate and useful, but calls out instances that cross the line. "*I* feel pain, my brain doesn't," Dennett insists. Other philosophers of mind, like Patricia Churchland and Derek Parfit, more fully embrace versions of the "you are your brain" view, based on their conceptions of what is most important about being "you." Parfit, for instance, associates personal identity with the experience of living an uninterrupted life story—what he calls "psychological continuity"—which depends most heavily on memories we think of as being stored in the brain.

But does understanding the relationship between our brains and our personhood extend beyond ivory tower philosophical hairsplitting? The great Wittgenstein is known for his declaration that philosophy arises from misunderstandings of language. Does the question of whether or not a person can be reduced to his or her brain just boil down to pedantic issues of how we *define* a person?

My answer is no, but I will make my points from a physiological perspective rather than a philosophical one. As we shall see, the brain interacts in essential ways with the rest of the body, and some of the most personal and individualized aspects of thinking and feeling depend critically on these interactions. If part of what makes you *you* includes your emotional side, your physical abilities, and the decisions you make, then it is scientifically inaccurate to equate yourself to your brain. Even the idea that your brain is in control of the rest of you is suspect, given that interactions between the brain and other organs tend to be reciprocal. By learning that the biological underpinnings of the mind have no sharp boundaries, we can more completely appreciate the integrated nature of mind, body, and environment, a crucial step in overcoming the cerebral mystique.

Our argument begins a long time ago, in a land far, far away. A young king lies lifeless on an embalming table, slowly dehydrating in the dry air of the land he recently ruled. An undertaker inserts a small chisel up the boy's left nostril and begins to tap at it. The chisel stops hard against the cribriform plate, a piece of bone separating the roof of the nose from the inside of the skull. A frisson ripples through the body with each tap, as if the poor lad is being repeatedly pinched in his postmortem dreams. Then with a final blow and a faintly audible crack, the bony plate gives way, and the chisel sinks deep into the dead boy's head. The undertaker extracts the tool, unstoppering a previously untapped reservoir of viscous liquid, which now begins to ooze out. Unperturbed, he picks up a small hook and thrusts it deep up the nostril where the chisel had been. For several minutes, the mortician wrenches and twists his hook through what seem like impossible angles, making mincemeat out of whatever is left inside the head. Then he begins drawing the hook in and out, easing out bits of grayish slime flecked with red. This goes on until there is nothing left to extract. The undertaker thrusts some wadding up the king's nose and declares the job finished.

Approximately so ended the brain of Pharaoh Tutankhamen, thirteenth ruler of the Eighteenth Dynasty in ancient Egypt's New Kingdom. In ancient Egyptian culture, preservation of the body and key organs was thought to be essential for well-being in the afterlife, but the brain was unimportant for this purpose. To the Egyptians, brains were merely "viscera of the skull," involved in such glamorous tasks as the production of mucus. If left in place after death, cerebral matter had a tendency to rot and spread decay; it was therefore unceremoniously extracted and discarded as part of the embalming process. According to the custom of the time, the actual viscera of Tutankhamen received considerably more respect than his brain. The king's stomach, intestines, liver, and lungs were carefully removed and preserved for eternity in funerary vessels called canopic jars. They were then entombed with the pharaoh's sarcophagus in the Valley of the Kings in 1323 BCE.

Ironically, if people today could choose a body part for use in the afterlife, the brain would probably be their first choice rather than their last. A group called the Brain Preservation Foundation, started by a former Harvard biology postdoc named Ken Hayworth, aims to promote "scientific research and services development in the field of whole brain preservation for long-term static storage"—a modern form of mummification, specific to the brain. The BPF offers a prize of more than $100,000 to researchers who can demonstrate techniques for effectively preserving human brain structure down to the synaptic level. Another organization, called Alcor Life Extension Foundation, maintains the frozen brains of about 150 "patients," who each paid tens of thousands of dollars to have their heads stored indefinitely in liquid nitrogen after their deaths. Kim Suozzi, whom we saw in the Introduction, is one of them. The hope is that eventually the technology will become available to thaw the brains and transplant them into new bodies, enabling a second life for the preserved brains' former owners. Alcor offers whole-body preservation as well, but this service is much more expensive and less popular. The brain may be the *only* necessary part of the body to preserve, argues Steve Bridge, a past president of the foundation. "We are our brains," he says, echoing voices we heard above.

But Tutankhamen's mummy suggests otherwise. Subjected to a battery of tests since its rediscovery by Howard Carter's archeological team in 1922, the brainless corpse tells us not only about the king's physical form but also about his mind. This is because the king was afflicted by bodily ailments that modern medical knowledge tells us must also have affected his character and experiences. For instance, X-ray CT analysis showed that the young pharaoh suffered from several manifestations of bone disease, including a crooked back, a clubfoot, and bone fragility that probably caused distracting pain. "He might be envisioned as a young but frail king who needed canes to walk," wrote the former Egyptian minister of antiquities Zahi Hawass and his colleagues in a landmark scientific analysis of the famous remains. DNA evidence from the mummy showed that the king also had severe malaria, a condition with well-established psychological consequences. Tutankhamen most likely experienced bouts of confusion and delirium that worsened as he approached his death.

The possibility of inferring aspects of a person's mental state from *extracerebral* bodily clues illustrates how the mind is intertwined with the body as a whole, and not only with the brain. We can speculate quite confidently about how a person with advanced malaria and bone disorders might feel, regardless of whether they had the brain of Einstein or the brain of Homer Simpson. Although the brain is required for a person's awareness of ailments, the routes by which disorders exert mental effects can be very indirect. The parasite that causes malaria, for instance, festers in the blood vessels and disrupts consciousness by interfering with blood flow and oxygenation, without ever entering the brain or encountering a neuron. Bone disorders affect the mind by causing inflammation and pain, mediated by parts of the nervous system far from the brain.

Conditions like these that inhabit the body's periphery alter the mind with regularity. We have all had the experience of feeling slightly dizzy when a bad fever hits us, but the influence of disease on mental state can run far deeper. Psychiatric complications can arise from maladies ranging from the common cold to cancer. As recently as the early twentieth century, the sexually transmitted bacterial disease syphilis carried more victims into insane asylums than many of the mental illnesses we know today. Syphilis starts with the development of unpleasant sores on the genitals, but left untreated can spread to organs throughout the body. Several years after infection, patients experience mood swings, psychotic delusions, and dementia, in a conjunction of symptoms known as neurosyphilis. The Romantic composer Robert Schumann is one of many eminent people (mostly men) thought to have died from this condition.

Schumann spent his last years in an insane asylum, and many critics have dismissed his later work as the product of a deranged or dilapidated mind. Others have found notable originality, however; musicologist Hans-Joachim Kreutzer sees evidence in the late work that "Schumann was always ahead of his times . . . [and] revolutionary in opening new musical worlds and developing them." Disease might have enabled some of Schumann's creative innovations just as hallucinogenic drugs can arouse the imagination of contemporary artists.

Psychiatrist Bradford Felker and his colleagues reviewed dozens of published medical studies in the 1990s and found that about 20 percent of psychiatric patients described in the papers had a nonpsychiatric, bodily medical problem—known in scientific terms as a *somatic disorder*—that might have produced or worsened their mental condition. Among these patients, mental dysfunctions ranging from depression to confusion and memory loss could have been caused by heart, lung, and endocrine disorders, or infectious diseases. Surprisingly, around half of the patients had no apparent knowledge of their underlying nonpsychiatric problems. This rules out the possibility that the patients' psychological conditions were induced by worrying about the somatic disorders. Instead, the somatic disorders most likely caused mental problems by altering blood sugar, oxygen supply, hormone balance, and a host of physical factors that couple the brain to the rest of the body. The possibility that one in five psychiatric patients could be misclassified and subsequently mistreated is alarming. But the broader principle demonstrated here is that the mental functions we usually attribute to the brain are actually functions of the body as a whole. When things go wrong with the body, the mind can suffer even if the brain itself is little more than collateral damage.

Does the influence of disease on the brain challenge the brain's position as the control center of the body? We know that the loftiest commander can be brought down by the lowliest of soldiers. King Harold Godwinson, defender of Britain against the Norman invasion, was felled in 1066 by an anonymous archer's shaft, shot through the eye at the Battle of Hastings. Czar Nicholas II of Russia was summarily executed by a former watchmaker from the Siberian hinterlands. Historical world leaders from the Roman emperor Caracalla to India's Prime Minister Indira Gandhi have been slain by their own menial bodyguards. Perhaps the effects of diseases on the brain could similarly be seen as rare and remarkable instances of usurpation by the subordinate against the supreme. We shall see in the next sections, however, that this is not the case. It is on a routine basis that the normal processes of behavior and cognition involve intimate interplay between the brain and body. The rest of the body unambiguously guides what we do, how we think, and who we are.

Nowhere is the integration of brain and body more apparent than in the domain of our emotions. Imagine coming home alone to your place one night to find the door unlatched and just slightly ajar. This shouldn't be—you locked it on your way out in the morning! Stepping gingerly inside, you wonder if you might have been burgled or, worse, if there is an intruder inside. Your pupils widen as you peer into the darkness of the room while groping for the light switch. You breathe faster, and your cheeks flush with blood. You find the switch and flick it on, stunning yourself with blinding light. At the same instant, a deafening roar of voices bursts forth in unison from the cavernous abyss in front of you. The muscles throughout your body seize up, your stomach clenches, and a fleeting faintness sweeps over you as your heart throttles into overdrive. Frozen in space and time, staring dumbly ahead, you are suddenly aware of an excited crowd of people in front of you, poised as if ready to spring. Your tunnel vision singles out just one face in the group. Most unexpectedly, it is the face of your college roommate, eyes wide and nostrils flared, his mouth just swallowing a cadence to the word "Surprise!" The tension dissipates as you remember that it's your birthday. You may be too old for surprise parties, but clearly your friends don't think so.

This is a scenario in which you are emphatically not only your brain. Your actions and sensations reflect the conjoined functionality of mind and body, and the integration of physiological processes that reach quite literally from your head to your toes. If you were an ancient human animal roaming the untamed savannah, the changes you underwent as you experienced fear of the dark unknown would have prepared you for the primeval fight-or-flight response. Your entire physique would enact whichever strategy you followed.

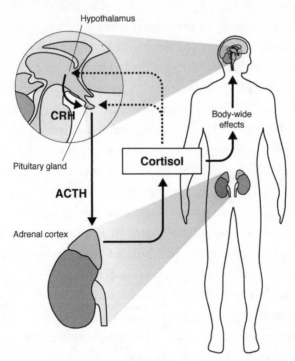

FIGURE 8. Diagram of the HPA axis showing interactions mediated by the hormones CRH, ACTH, and cortisol (negative feedback interactions indicated by dotted arrows).

The biology behind your flushing cheeks, your racing heart, your tensed muscles, and your tunnel vision is closely related to a network of structures called the *hypothalamic-pituitary-adrenal axis*, or HPA axis (see Figure 8). The hypothalamus is a brain region known for its secretion of neuropeptides and hormones, components of the complex chemical soup discussed in Chapter 2. One of the hypothalamic hormones, a molecule called *corticotropin releasing hormone*, or CRH, is released into the bloodstream during stress and quickly reaches the pituitary gland, a pea-sized hormone production factory that sits just below the brain. Pituitary cells stimulated by CRH secrete a second substance called *adreno-corticotropic hormone* (ACTH) into the bloodstream. ACTH acts on the adrenal glands, a pair of yellowish blobs that sit on top of the kidneys, inducing them to release a third hormone known as *cortisol*, which induces increases in blood pressure and metabolism throughout the body. In parallel with this chemical signaling pathway, a neural tract leads directly from the hypothalamus to the adrenal glands; this tract is also activated during stress and results in the release of adrenaline, another small hormone that synergizes with cortisol to induce changes such as increased heart rate and blood flow.

Although the HPA axis clearly extends well beyond the brain, it appears superficially like another example of top-down control until one considers how bodily input and feedback influence how the system works. Consider for instance how your visual perception changes as your anxiety increases. The pupil dilation you experience is driven largely by adrenaline. Wider pupils help you see sharply but also lead to the loss of peripheral acuity, explaining your tendency to focus on a single object (your roommate's face) without taking in the full scene. Adrenaline and cortisol also contribute to your perception of danger by creating the somatic symptoms you associate with stress. Consciously or subconsciously, you feel the pace of your breathing, the wrenching of your gut, and the warmth of blood engorging your face and muscles. This creates a feedback loop between the brain and the rest of your body in which each helps drive the other further into red alert. Fortunately, this vicious cycle is counterbalanced by an opposing brain-body feedback loop in which cortisol from the adrenal gland suppresses the production of CRH and ACTH in the hypothalamus and pituitary, providing direct chemical signals from the body to the brain that tend to keep the system under control.

The extent to which the brain appears to be the prime mover of emotional responses is highly variable. In the case of the stressful first scene of the surprise party, your anxiety was triggered by the suspicious circumstances you found when you got home. Your realization that something was wrong depended on your memories, stored for the most part in neural structures, even if the consequences were played out by body-wide physiology. In other instances, however, stress and anxiety can be triggered by factors outside the head. A classic example occurs in pregnancy. During gestation, the placenta acts as an abnormal source of extra CRH, leading to ever-increasing levels of cortisol in the mother's bloodstream that cannot be governed by the normal feedback loops. Once the baby is born and the placenta removed, a sudden drop in cortisol occurs. These hormonal changes and their effects on the HPA axis help drive the dramatic mood swings many women experience before and after childbirth.

All emotions—not just anxiety—are linked to extracerebral bodily changes and bodily sensations. We say that our guts are wrenched by sorrow, our hearts throb with love, our blood boils in anger, and so on; although these phrases are largely figurative, they also reflect underlying physiological reality. No less than Charles Darwin was one of the first to make an extensive study of the relationship between emotional states and bodily changes. Drawing on extensive observations from humans and animals, Darwin concluded that emotional states are reflexively expressed through physical actions that in evolutionary history served to relieve or gratify the emotions. An angry man, for instance, "unconsciously throws himself into an attitude ready for attacking or striking his enemy," while people describing a horrific sight might "shut their eyes . . . or shake their heads, as if not to see or to drive away something disagreeable." William James, one of the fathers of modern psychology, carried the notion of emotional-physical coupling yet further. He theorized in 1890 that the complex of bodily changes during an emotional experience doesn't just express the emotion—it *is* the emotion. "We feel sorry because we cry, angry because we strike, afraid because we tremble," James wrote, rather than "that we cry, strike, or tremble, because we are sorry, angry, or fearful."

More recent studies have refined nineteenth-century ideas by using biomedical measurement methods to connect emotions more precisely to bodily phenomena such as muscular activity, heart and breathing rates, and skin conductance changes like those used in lie detection tests. In a survey of over a hundred such studies, Sylvia Kreibig of Stanford University saw evidence for specific relationships between multiple types of emotions and bodily responses they correlate with. Even relatively similar emotions could be distinguished according to physiological differences; for instance, although anxiety and sadness (without crying) both incur increased breathing rate, the two emotions tend to go with opposite changes in heart rate, skin conductance, and respiratory volume.

A fascinating 2014 analysis by Lauri Nummenmaa and colleagues at Aalto University gave more corporeal form to emotional responses by generating "bodily maps" of the affective experiences of over seven hundred

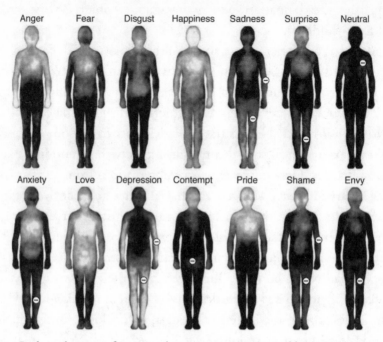

FIGURE 9. Body-wide maps of emotional perception, as measured by Nummenmaa and colleagues. Circles with minus signs denote negative response areas. Adapted with permission from L. Nummenmaa, E. Glerean, R. Hari, and J. K. Hietanen, "Bodily maps of emotions," *Proceedings of the National Academy of Sciences* 111 (2014): 646–651.

subjects. Participants used diagrams of the human body to report the sensations they associated with fourteen emotional or neutral states. By averaging input from all the participants, the scientists found distinct body-wide patterns of perceived activation or deactivation corresponding to each emotion (see Figure 9). A feeling of sadness in this study went with diminished sensations or deactivation in the limbs and extremities, while a feeling of love went with strongly heightened sensations in the face, upper abdomen, and groin. To eliminate linguistic or cultural biases, the researchers performed parallel studies on Finnish, Swedish, and Taiwanese subjects; they found similar results among the groups. In addition, the researchers discovered a close correspondence between the bodily effects subjects consciously associated with each emotion and the sensations that were *actually* evoked by emotional videos or stories.

Body-wide emotional responses like the ones Nummenmaa mapped may play an integral part in cognition. In the mid-1990s, the psychologist

Antonio Damasio articulated an influential theory about how emotional sensations influence the process of making decisions. Damasio argues that we subconsciously learn to associate our decisions with the emotional bodily changes our actions lead to. He refers to these learned bodily associations as *somatic markers*. Somatic markers are recalled when we face choices similar to what we have experienced in the past, and they provide instantaneous feedback that either favors or disfavors each available course of action. "When a negative somatic marker is juxtaposed to a particular future outcome the combination functions as an alarm bell," Damasio writes. "When a positive somatic marker is juxtaposed instead, it becomes a beacon of incentive." I might be trying to decide between eating lunch in a restaurant downtown and one near where I live, for instance. The downtown restaurant has better food, but I would need to take the subway to get there, enduring crowds, darkness, noise, and the smell of urine. Negative somatic markers associated with the subway are activated (perhaps slowed breathing, tightening of the throat), and I decide in a flash to choose the option near my house. In Damasio's framework, the bodily responses help me make this decision without thinking through all the details, providing a shortcut that links the possible course of action to its likely emotional outcome.

The motivation for Damasio's somatic marker hypothesis came from his observations of patients with damage to a brain area called the *ventromedial prefrontal cortex* (vmPFC). This brain region seems to be activated by sensory input from the body, and it could be an interface between sensory signals and higher cognitive functions of the brain. Damasio notes that patients with vmPFC lesions also have "compromised ability to express emotion and to experience feelings in situations in which emotions would normally have been expected." Notably, these patients perform respectably on IQ tests but make bad decisions in situations that involve risks or long-term outcomes such as social relationships or business deals. The poor performance of vmPFC patients in many strategic situations also contrasts with the stereotypical view, personified by *Star Trek*'s emotionless but brilliant Mr. Spock, that emotions are opposed to reason and counterproductive to human problem-solving.

Although some critics question the role of somatic markers per se in decision making, Damasio's more general point about the interplay between brain and body in emotion and cognition is becoming widely accepted. "The opportunities for bodily feedback during emotional reactions to influence information processing by the brain and the way we consciously feel are enormous," writes neurobiologist Joseph LeDoux in his book *The Emotional Brain.* In LeDoux's telling, emotions provide an evolutionarily tuned mechanism by which animals can make rapid, almost reflexive decisions that tend to keep them happy and out of trouble. Such a mechanism is also at play in human economic behavior, according to Princeton University psychologist and economist Daniel Kahneman. Summarizing results of extensive psychological experiments he performed with Amos Tversky in the 1970s and '80s, Kahneman explains that choices are often guided by a mental process that "operates automatically and quickly, with little or no effort and no sense of voluntary control." This fast decision process "effortlessly originat[es] impressions and feelings that are the main sources of the explicit beliefs and deliberate choices" executed by slower mental operations. In other words, emotionally guided decision making either directly or indirectly influences most of what we do.

At an even more fundamental level, the bodies that contain our brains influence our minds through the physical possibilities they give us. It is no secret that virtually all of our actions depend on the capabilities of our bodies, but the extent to which even the most brainy activities are shaped by relatively mundane properties of our bodies can be astounding. The legendary violinist and composer Niccolò Paganini is thought to have suffered a connective tissue disorder that gave his hands abnormal flexibility. "His hand was not larger than the proper size but he doubles its width due to the elasticity of all the structures within it," recounted his personal physician Francesco Bennati in 1831. No amount of focus, determination, or abstract thinking could thus have given Paganini the advantage he got from his special joints. And the musician's most cerebral acts—his own original compositions—followed just as surely from his unusual physical characteristics. He designed fabulously virtuosic pieces that few other than he could play, using harmonics to achieve tremendous range on single violin strings, simultaneous bowing and pizzicato, and chords played on all four strings at once. What Paganini accomplished and in a deep sense *who he was* were thus inseparable from his physique. "To become Paganini, musical genius was not sufficient: one needs the physical structure which he holds," wrote Bennati.

There is probably no more intellectual activity than the discipline of pure mathematics, but even the way mathematicians think may involve their bodies. Is it a coincidence that many of the Frisbee-tossing, cafeteria-tray-spinning, beanbag-juggling geeks from my high school and college years became mathematicians and physicists in adulthood? A certain type of physical activity—one involving complex bodily manipulations of three-dimensional objects in space—seems to be part of the milieu from which highly analytical minds can emerge. Even Carl Gauss, the legendary mathematician whose brain we met in Chapter 1, was said to have scaled three mountains in central Germany in a supremely physical effort to test ideas about triangle geometry. According to linguist George Lakoff and psychologist Rafael Núñez, the relationship between mathematics and physicality is more than superficial. They see mathematical thought as a phenomenon rooted in human sensorimotor experience, as opposed to the pure appreciation of objective truth. Lakoff and Núñez write that "mathematics is the product of the neural capacities of our brains, the nature of our bodies, our evolution, our environment, and our long social and cultural history."

Scientists and philosophers use the term *embodied cognition* to describe the view that cognitive processes emanate from entire physical organisms and their interactions with the world, rather than just their brains. A particular emphasis of the embodied cognition movement is on how gross features like body plan and spatial experiences lead to ways of thinking and behaving. "Our bodies and their perceptually guided motions through the world do much of the work required to achieve our goals," write psychologists Andrew Wilson and Sabrina Golonka, suggesting that this replaces the need for more abstract forms of cognition. "This simple fact utterly changes our idea of what 'cognition' involves," they say.

Embodied cognition is on display as a colony of beavers builds a dam. The animals begin by driving branches into the stream bed and weighting them down with stones. They build up from this base, expanding the construction with wood and debris. The structure is filled in with more stones, bark, moss, sticks, and mud dredged from the streambed. Beaver dams are usually about a meter wide and about two meters tall, and sculpted in form to match the water's depth and flow rate. At first glance, there is little to distinguish the beaver's product from what an MIT-trained engineer working with the same materials might create—there is a reason why the beaver is MIT's mascot! But now put the beavers themselves back into the picture. Each beaver has sharp, iron-reinforced hatchets for teeth, a mason's trowel for a tail, the webbed feet of an Aquaman, and the low frame and powerful limbs of a little stevedore. If you were to design an animal whose job it was to dam streams, there's a good chance you'd come up with a beaver. The all-important and apparently cognitive decision to get into the dam-building business in the first place is also hardwired into the beaver. The sound of running water gets them so exercised that they will obsessively work to stop it by building or repairing their structures. It is an instinct that has been ingeniously exploited by hunters in search of beaver pelts; they sabotage the dams and lay traps for the poor animals that show up in response.

There are many other instances of embodied cognition in nature, but some of the best examples come from the field of robotics. In her book *Beyond the Brain*, psychologist Louise Barrett gives the example of the "Swiss robots" who run around collecting objects and clustering them together as if they were tidying up. This apparently complex behavior is not programmed into the robots—it follows from the basic principles of their design. The robots are equipped with a set of proximity sensors that allow them to detect objects off on their sides, but not in front of them. As they move, the robots also have to obey a single instruction: if a sensor on one side is activated, turn in the opposite direction. Because each robot doesn't care about objects in front of it, it will tend to run into things and push them along, but if it detects a second object off on the side as it is doing this, the robot will turn away, leaving the first object together with the second one in a small cluster; and this process continues indefinitely. The Swiss robot's decluttering behavior provides a good example of how simple physical capacities like the abilities to push, sense, and turn can give rise to sophisticated behavior without any planning or top-down cognitive control.

The design of the human body determines our behaviors as well. If you have a dog, you might at times have wished you could tell her to go walk herself, feed herself, or take care of herself in other ways. If only your dog were smarter, you could just explain these things to her and let her be self-sufficient while you go to work, take a vacation, or whatever. But the truth is that if your dog suddenly acquired human intelligence and the ability to speak English, chances are overwhelming that she would still remain more or less as helpless as a baby. She wouldn't be able to open most doors on her own, she might not be able to turn the faucet on and off, she probably couldn't get food ready for herself, and she certainly wouldn't be able to go grocery shopping. This is because our environments are constructed to provide behavioral possibilities only we humans can take advantage of—what the psychologist James Gibson termed *affordances*. Our doorknobs, computer keyboards, chairs, and beds all offer affordances matched to our anatomy. Without our bipedal stance, binocular vision, fine motor capabilities, and ability to grasp, almost all of the basic activities we indulge in would be literally out of reach.

George Lakoff and his colleague Mark Johnson have postulated that our bodies ultimately determine not only what we do but also how we think. In their classic *Metaphors We Live By*, Lakoff and Johnson argue that most of the concepts we use in language or thought are built up from simpler ideas by way of metaphors. "Because so many of the concepts that are important to us are either abstract or not clearly delineated in our experience (the emotions, ideas, time, etc.)," they write, "we need to get a grasp on them by means of other concepts that we understand in clearer terms (spatial orientations, objects, etc.)." In Lakoff and Johnson's view, our conceptual systems are deeply rooted in the experiences we get from moving through the world with our particular bodies—in other words, even our higher cognition is embodied. We don't need help understanding concepts like "up," "have," or "push," because they are so closely related to physical fundamentals like the way we stand and what we can do with our hands. To show how these fundamentals provide building blocks for more complicated ideas, Lakoff and Johnson point out that inherently nonspatial or nonphysical concepts are nevertheless discussed in spatial and physical terms. When we speak about happiness, for instance, we use language implying that "happy is up"; we say our mood rises or our spirits are high. When we speak about time, we treat it as a resource we can seize, hold, or gather; we can have or lose time, or we can give someone a portion of our time. When we speak about having an argument, we use language that refers to a physical fight; we oppose each other, and our reasoning may be either strong or weak.

There is interesting experimental evidence for the Lakoff-Johnson view from studies that relate people's physical posture to their thoughts and attitudes. In one example, Anita Eerland and colleagues at Erasmus University tested the idea that thinking about quantities has to do with how people mentally visualize numbers on a number line, and they explored how this might in turn be linked to body position. The researchers told subjects to stand on an imperceptibly sloping ramp and asked them to estimate various numerical quantities, like the weight of an average elephant or the height of the Eiffel Tower. When the ramp sloped left, the subjects leaned left, as if they were tilted toward the low numbers on a scale or meter. Amazingly, this subtle change in stance correlated systematically with lower answers to the questions. It was as if pachyderms were lighter and Alexandre Eiffel's edifice was smaller to left-leaning participants than to others who stood straight or leaned right. In another demonstration of the mind-posture relationship, a team led by Lynden Miles at the University of Aberdeen attached motion sensors to twenty volunteers and then asked them to think about events in the past or events in the future. In this case, the subjects leaned forward when thinking about the future, but backward when thinking about the past.

Some of the strongest evidence for a connection between physicality and cognition comes from studies of the effects of exercise on mental functions. These studies support the general conclusion that an active body promotes an active mind. Perhaps the greatest interest in this idea stems from the enticing suggestion that exercise programs might stave off cognitive decline during aging. In adults over fifty, regular forty-five- to sixty-minute sessions of moderate to vigorous exertion do in fact increase performance in tests of memory, attention, and problem solving; improvements occur both in healthy subjects and in people with the mild cognitive impairments seen in early-stage Alzheimer's disease. There is some evidence for cognitive benefits of exercise in younger people as well. In one of the most impressive demonstrations, psychologists Marily Opezzo and Daniel Schwartz of Stanford University had college students sit on a chair or walk on a treadmill in an empty room and then undergo standardized tests of creative thinking. Compared with sedentary participants, the people who had been walking on average gave better answers when asked to come up with novel analogies or think of unusual uses for common objects. In a related experiment, subjects either walked through campus or were pushed along the same route in a wheelchair; again the students who had walked seemed to get a more pronounced cognitive boost, as measured by creativity tests.

We may be tacitly evoking the relationship between physical activity and cognition when we say that someone has an "agile mind," or that she is "thinking on her feet." But this particular relationship goes well beyond metaphors. Exercise is now known to produce physiological changes that directly couple the brain to the rest of the body. Over a short term during and immediately following exertion, the heart pumps faster, and the brain experiences increased blood flow, which increases its oxygen and energy supply. The brain also produces locally acting chemical substances called *neurotrophic factors*, which promote cell growth and maintenance. Over the longer term, these chemicals induce the formation of new neurons in an area of the brain called the *hippocampus*. This area is specifically important in memory formation, and the replenishment of its cells may be particularly critical for counteracting the effects of Alzheimer's disease, which causes cell death in this brain region. The effect of exercise on cognition thus shows that relatively superficial things we do with our bodies can alter our minds through defined physiological pathways like those we considered in the previous section.

We have now looked at multiple ways in which biology beyond the brain contributes to thinking and feeling. Clearly we would not be the people we are without our physical attributes, without our health, or without physiological mechanisms like the HPA axis. In cases like the composer Paganini or the dam-building beavers, we see that bodily characteristics help determine the mental activities each individual is capable of, and in examples like the stressful surprise party or the phenomenon of cognitive enhancement through exercise, we can see how brain and body interact in a closed loop of physiological interdependence. But a particularly convincing demonstration of the importance of the body on the mind might come from a simplistic experiment: transplant a new brain into a person's body and see what characteristics they inherit from the brain versus the body. Unfortunately the brain transplant is beyond our current capabilities, but a less radical procedure—the transplantation or alteration of parts of the body—is both feasible and widespread. Can organ and tissue transplants change a person's mind or personality?

A growing amount of evidence suggests that the answer is yes. The most sensational stories come from people like heart/lung transplant recipient Claire Sylvia. In her book *A Change of Heart*, Sylvia recounts how she inherited not only the organs but also the personality and memories of her donor. She traces her awareness of this to a night after her surgery, when she experienced a dream in which she met and kissed the donor, Tim L. "I woke up knowing—really knowing," Sylvia writes, "that Tim L. was my donor and that some parts of his spirit and personality were now in me." A recent article in the London *Telegraph* tells of Kevin Mashford, who became a bicycle enthusiast after receiving the heart of a cyclist killed in a crash, and Cheryl Johnson, who thought she had gained a highbrow taste in literature from her kidney donor. Alluring as such accounts are, there is no scientific basis to the idea that organs outside the brain store or transmit specific memories or proclivities. At the same time, science does not dispute the sheer fact that these patients experienced extraordinary psychological upheavals as a result of their surgery.

There are good reasons to believe that transplants such as Claire Sylvia's can broadly affect the mind. "Heart transplants trigger a number of significant physiological and psychological changes, and the overall result depends on the individual," writes *Slate* technology editor Will Oremus. "One of the most common effects is straightforward: A new lease on life tends to make people happier and more optimistic." A team of Austrian researchers performed one of the first substantial psychological studies on transplant patients by interviewing forty-seven heart recipients at a hospital in Vienna in 1992. About 20 percent of the patients reported personality changes within two years of their operations. Most of these subjects attributed alterations to the near-death experiences they had leading up to their operations, but some did not. Those most affected reported changed preferences or worldview. "The person whose heart I got was a calm person, not hectic," speculated one of the patients, "and his feelings have been passed on to me now."

The heart symbolizes love and passion, and perhaps these associations color the reports of heart transplant patients. What about replacement of less glamorous organs? Not surprisingly, given everything transplant patients often go through, cognitive and emotional alterations are observed almost across the board. Sometimes these can be traced to specific biochemical or hormonal pathways by which the peripheral organs affect the mind. The most straightforward changes arise from correcting the organ failure that brought about the need for a transplant in the first place. For instance, both kidney and liver failure increase the levels of toxic substances like urea or ammonia in the blood, a situation that can produce personality changes and intellectual difficulties. Replacing the diseased organ with a healthy one resolves these problems and can directly benefit cognition. A rare before/after comparison among twelve patients in Italy showed in fact that liver transplantation produced long-lasting mental improvements across a range of domains, in tasks related to attention, memory, and spatial problem solving.

Medical manipulations of the digestive system seem to exert particularly strong effects on how people think and feel. This special relationship probably owes much to an extensive network of biochemical and nervous communication channels that connect the brain and the gastrointestinal tract. Major components of the network include the *vagus nerve*, a branched neural superhighway that links most of the abdominal organs to the brainstem, and the *enteric nervous system*, a largely independent "second brain" of more than a hundred million neurons in the gut that interacts bidirectionally with the brain in our heads. Invasive changes to the digestive system can be expected to impinge either directly or indirectly on such structures.

The most common surgical procedure that alters the digestive organs is not a transplant but rather a partial removal of the stomach; this is a relatively common treatment for obesity. It turns out that around half of weight-loss surgery patients experience personality shifts, according to the advocacy website ObesityCoverage.com. Electronic message boards related to these procedures are filled with accounts of Dr. Jekyll to Mr. Hyde transformations, or occasionally the reverse. "I am bombarded with new and wonderful emotions," writes one bariatric patient. "I love who I am now, but I'm really confused. I am a person I have never been—even when I was thin before and young." The prevalence of such disruptions may help explain why stomach surgery is followed by a high incidence of divorce. Are dramatic life changes due to physiological consequences of stomach alteration itself, or to repercussions of the patients' subsequent weight loss? These factors are nearly impossible to tease apart, but both are likely to contribute.

One of the most bizarre examples of transplantation modifying mental makeup also comes from the digestive system. In this case, what is transplanted is not an organ but rather the community of bacteria and other microorganisms that live in the intestines—the so-called *gut microbiome*. Gut microbiome changes can be produced either through a somewhat offputting procedure called a fecal transplant (think about it) or by having subjects ingest bacteria-rich material like yogurt. Bacteriotherapy methods like these are used to treat intestinal infections by harmful bacteria like *Clostridium difficile*, which causes colitis in many thousands of US patients each year. The procedures increase the proportion of "good bacteria" that normally inhabit healthy guts; the beneficial organisms can then reduce inflammation by literally crowding out the toxic microbes. And recent research supports the startling conclusion that such changes in the makeup of the gut microbiome affect not only digestive functions of the intestine but also psychological states like anxiety, stress, and depression.

Bacteriotherapy is increasingly applied in people, but the most striking behavioral results have come from laboratory studies with mice. In one experiment, a team led by Stephen Collins of McMaster University performed two-way fecal transplants between mice of two strains, called BALB/c and NIH Swiss. Normally BALB/c mice are more timid than NIH Swiss mice; they tend to avoid light and stay put in one place rather than exploring their surroundings. Amazingly, after BALB/c mice received the fecal matter from NIH Swiss mice, they became more outgoing and exploratory. Conversely, when NIH Swiss mice got the stuff from BALB/c animals, they became less exploratory and more anxious. In a different experiment, a team led by John Cryan at the University of Cork fed a group of mice with broth containing *Lactobacillus rhamnosus*, a "good bacterium" found in some dairy products. The mice became more resilient and less sensitive to stress than animals that didn't get the special drink—they explored more, they were less easily startled, and when dumped in a bucket of water, they swam for longer without giving up. Both the Collins and the Cryan gut microbiome manipulations produced neurochemical changes in the brain, and fMRI studies in humans have shown a relationship between bacteriotherapy and brain responses as well. The physiological network that enables microorganisms in the gut to influence brain processes is now referred to as the microbiome-gut-brain axis.

These tales of mice and men show repeatedly that transmutation of the body can affect the mind, changing the way individuals feel and behave in surprising and significant ways. The same person with the same brain, after a transplant or therapy that affects the rest of their body, can find themselves with an altered personality or outlook. In this we can find a remarkable counterpoint to one of the most celebrated case studies in the history of neuroscience: the story of Phineas Gage, a rail worker who in 1848 miraculously survived a freak explosion that blew a hole through part of the prefrontal cortex of his brain. After his injury, the victim became more impulsive, reckless, and crude, damaging his relationships and prospects for employment. Gage was "no longer Gage," witnesses reported at the time, observing some of the same kinds of emotional and decision-making problems that Antonio Damasio worked into his somatic marker hypothesis. In this section we have seen that a lifesaving transplant, stomach stitching, or gut microbiome makeover can bring about personality changes that seem comparable in magnitude to what Gage underwent—but without touching the brain.

I was recently at a meeting with US congressional staffers who had come to my university to learn about neuroscience research going on there. After some brief presentations from the academics, there was time for a few questions from the visitors. The hands shot up. One of the first people called upon got straight to the point: "Will cognitive enhancement be possible?"

This question instantly calls up images of postapocalyptic beings studded with brain prosthetics, like the dreaded Borg of *Star Trek*. Slightly friendlier examples include much lower-profile technological implants that enhance the memory and sensory capabilities of characters in the Canadian science fiction series *Continuum*. Advances in brain-machine interface technology make the possibility of cognitive enhancements via such invasive mechanisms seem more realistic. Over thirty thousand people worldwide have worn brain implants that help alleviate movement disorders like Parkinson's disease and dystonia by delivering tiny electrical stimuli to specific brain regions. The development of optogenetics, the optical neurostimulation method we saw briefly in Chapter 2, offers hope that scientists might one day be able to manipulate the brain as easily as one types on a computer keyboard—inputting new information, controlling attention, and performing calculations faster than any natural brain could ever do. It is both reasonable and timely to ask how close such fantasies are to reality.

But my colleague Laura Schultz was thinking out of the box when she answered the questioner with another question: "Did you have a cup of coffee this morning?" What Laura was getting at was, of course, that coffee is a low-tech cognitive enhancement many of us use on a daily basis, and that doesn't involve anything like the futuristic brain implants or surgical interventions the questioner probably had in mind. Although coffee is not as cool as cyborg gear, it's much easier to come by and good to the last drop while working on a tough project.

Laura's answer was outside the box because it was a clever departure from stereotypical thinking about cognitive enhancement. Her answer also evokes this chapter's theme that the factors influencing our cognition do not necessarily go directly "into the box" and act exclusively on the brain. Coffee is innocently ingested via the digestive system. The active ingredient in coffee, caffeine, spreads throughout the body and produces changes both in the brain and beyond. It directly interferes with a chemical signaling molecule called *adenosine*. Blocking adenosine in a part of the brain called the *ventrolateral preoptic area* contributes most directly to arousal, in turn leading to a pattern of body-wide changes that

include increased blood pressure and cortisol levels, as well as mild anxiety or stress. Caffeine therefore recruits the integrated interconnections
between brain and body, and the subjective feelings of alertness we get
from a cup of coffee depend on this.

In Part 2 of this book, we will take a closer look at how the cerebral
mystique influences how we view drugs and technology that affects our
minds by acting inside or outside the brain. But first we will explore one
last and particularly important facet of the mystique itself: the idea that
our brains and minds can be separated from the environment around us.

six

NO BRAIN IS AN ISLAND

"LET US MAKE MANKIND IN OUR IMAGE, AFTER OUR likeness," says the God of Genesis, "and let them have dominion over the fish of the sea, and over the fowl of the air, and over the cattle, and over all the earth, and over every creeping thing that creepeth upon the earth" (Genesis 1:26). Except for the Lord himself, humanity reigns supreme over the planet, destined to shape rather than be shaped, command rather than be commanded, and consume rather than be consumed. A similar belief in the hegemony of *Homo sapiens* is shared today by most people of the world, regardless of their religion or lifestyle. Even committed vegans or Buddhists, who ostensibly reject the subjugation of animals, participate in a global civilization that progressively subdues the planet through urbanization, transportation, agriculture, and industry. We are, as many now say, living in the age of the Anthropocene.

It is an understatement to assert that the triumph of humankind over nature could not have been possible without our brains. We each seem to control our immediate environments through our ability to perform voluntary actions, a capability once associated with the metaphysical soul or self but more recently swallowed up by the central nervous system. "Increasingly it is recognized that the self does not need to be outside the body," wrote renowned neuroscientist Peter Milner in his 1999 book *The Autonomous Brain*. "It may more conveniently be thought of as a complex neural mechanism, often referred to as the brain's executive system." University College London cognitive scientist Patrick Haggard observes likewise that "modern neuroscience is shifting towards a view of voluntary action being based on specific brain processes."

Part of the blame for this shift goes to the famous experiments of people like Benjamin Libet, who showed in the 1980s that electrical signals from the brain can be used to predict a person's apparently voluntary movements even before the person has made a conscious decision to act. In effect, the brain "knows" what we will do before we ourselves can catch on. This suggests that the brain is boss, potentially determining our actions and having our ideas more or less on its own. As neuroscientist David Eagleman puts it, the brain is "the mission control center that drives the whole operation, gathering dispatches through small portals in the armored bunker of the skull."

But the brain-in-a-bunker metaphor presents a paradoxical view of the interaction between our brains and their environments. In addition to understating the connections between brain and body, this picture presents the wider environment as little more than a passive contributor to brain function. The environment supplies information to the bastioned brain, which then reviews the data and decides how to respond. The brain is the commander, and the originator of all autonomous action. The thickness of the bunker's armor and the smallness of its portals imply that the commanding officer is well insulated from the action in the field, left to ponder intelligence and plan strategy far away from the actual fighting. But in the stark distinction between what goes on outside versus inside the brain, and in the unbalanced power relationship between them, we can see another example of the scientific dualism we have considered in previous chapters of this book.

The paradox here lies in the difficulty of identifying a place where dispatches from the outside turn into decisions. Can we conceive of a hand-off point—either a distinct locus or perhaps a more distributed "complex neural mechanism"—where the deterministic response to environmental

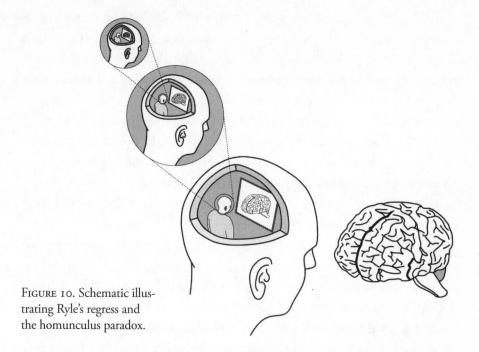

FIGURE 10. Schematic illustrating Ryle's regress and the homunculus paradox.

input ends and the brain's cognitive control kicks in? Some philosophers have argued that having such a place is equivalent to supposing that a little person—a *homunculus*—sits in the brain, receiving all of the input and figuring out how to respond. The 2015 Disney hit *Inside Out* spins this cartoonish scenario into a literal cartoon, in which five personified emotions named Joy, Sadness, Fear, Anger, and Disgust vie to influence a young girl's actions by twiddling knobs and levers on a control panel inside her head. But how do Joy and her colleagues accomplish this? Doesn't there need to be another set of homunculi inside each of them, converting *their* input to output? This notion applies recursively, like the never-ending cascade of reflections you see when you stand between two opposing mirrors in a dressing room (see Figure 10). The resulting contradiction is known as "Ryle's regress," after the philosopher Gilbert Ryle, who discussed the problem in his 1949 magnum opus, *The Concept of Mind*.

An obvious way to avoid Ryle's regress is to give up on the idea that the brain can ever operate independently of the world around us. In this view, the decisive influence of the environment reaches much more deeply into the brain, all the way to our decisions and actions themselves. When an apple drops from a tree, when the winter snows melt into the rivulets of spring, or when a motorcycle skids off the highway into a ditch, it is the laws of physics and the contours of the surroundings that determine their paths. Perhaps the brain is like a falling apple, governed by the forces of nature even as it strikes Isaac Newton on the head. In his famous essay *On the Freedom of the Will*, the early nineteenth-century philosopher Arthur Schopenhauer insisted that "like the actions of every being in nature, [human doings] must be subject to the law of causality in all its strictness." If this were true, the human brain would simply be an element in the causal chain, a bead vibrating passively on the string of life, rather than a hand that shakes the string itself. Nature would have dominion over the brain, rather than the other way around.

A difference between our time and Schopenhauer's is that today we have access to a wealth of experimental data about the long arm of the environment at work in our brains and in our behavior. In this chapter, we will survey some of this evidence and see that the causal role of the environment is more than just a theoretical abstraction. The relationship between brain and environment goes much further than the truisms that people are products of their time and place, that both nature and nurture are important, or that learning and memory arise from experience. When the boundary between the brain and environment is blurred, every thought and action, even at the moment it occurs, becomes a consequence of influences in the wider world. By examining these connections, we will again combat the mystique of the brain as controller and appreciate the extent to which our brains are natural entities subject to the universal laws of cause and effect.

A famous allegory of the interaction between the environment and the mind is the statue of the three wise monkeys. The oldest example is said to be a seventeenth-century carving over a doorway at the Tosho-gu shrine, near the burial place of the great shogun Tokugawa Ieyasu in Nikko, Japan. One monkey covers his eyes with his hands, another clutches his ears, and the third muzzles his mouth, representing the timeless injunction to "see no evil, hear no evil, and speak no evil." Monkeys are famous for getting around, and these three are no exception. Figurines of the monkeys are now among the world's most globalized forms of kitsch, sold across six continents, with a small but dedicated society of collectors that runs a website and meets annually. A statuette of the monkeys was one of Mahatma Gandhi's few possessions, symbolizing his strict moral code. Elements of the Italian mafia took the monkeys to represent their own code, the *omerta* or vow of silence. Some depictions humorously promote chastity by adding a fourth "do no evil" monkey who guards his private parts. The animals' input and output are both covered in this way. By juxtaposing seeing and hearing with saying and doing, the monkeys thus teach the lesson that behavior is inseparable from external influences that begin with the senses.

Our principal sensory systems—sight, hearing, touch, taste, and smell—offer the clearest routes by which the environment around us influences our thoughts and actions. It is telling that another Eastern allegory, an ancient Hindu version of Plato's famous mind-chariot analogy, presents the senses as five horses pulling a vehicle symbolizing the body. Almost all of the things we learn are ingested via our sensory organs, but the senses provide much more than fodder for our education. Sensory systems allow environmental stimuli to shape our thoughts and actions in far more immediate ways. On a continuous basis, we are inundated by the steady stream of input from our sensory organs to the brain. Like a drink from the proverbial fire hose, the force of this information flow is both overwhelming and unstoppable. Our senses remain active even in sleep and under anesthesia, relaying signals into the brain whether we are consciously aware of them or not. By delving into sensory biology, we can begin to appreciate how difficult it is for our brains to resist the influence of external stimuli.

The best studied and probably most influential of our senses is sight. Vision researchers have spent many decades studying how light striking the photosensitive part of the eye, the retina, is detected by the so-called rod and cone *photoreceptor cells* and subsequently processed into neural impulses (action potentials) that zip toward the brain via the optic nerve. By measuring electrical signals from the retinal output neurons called *ganglion cells*, neurophysiologist Horace Barlow discovered in the 1970s that each individual quantum of light—each photon—can give rise to an average of one to three action potentials. Even in complete darkness, Barlow found that some ganglion cells can fire as many as twenty of these spikes per second; this activity is essentially noise in the system, but it still floods the brain's inbox.

Caltech professor Markus Meister devised a way to analyze the first steps in vision by isolating live animal retinae and spreading them like tiny blankets onto beds of recording electrodes, permitting measurements from dozens of ganglion cells simultaneously. Using this technique, Meister and other neuroscientists have observed how retinas adapt rapidly to huge changes in image intensity and contrast, ensuring that the deluge of visual information to the brain is never stanched. In one study, researchers estimated that total data transmission from the human eye to the brain is roughly equivalent to a computer's internet connection, moving about a megabyte of visual input (four million spikes) each second over neural wires formed by the axons of a million ganglion cells per retina.

The nonvisual senses are also prolific sources of input to our brains. The *organ of Corti* in the inner ear converts sound waves into neural impulses—it is the ear's equivalent to the retina. Most of the auditory neurons that emerge from the organ of Corti fire action potentials at a rate of more than fifty spikes per second, even for low sound levels. With about thirty thousand auditory neurons per ear in humans, the total number of action potentials reaching the brain from the ears each second extends well into the millions. An enormous amount of input also comes from the body's largest sensory organ, the skin. Normal skin contains four types of touch or pressure-sensitive receptor cells, two types of temperature-sensitive receptors, and two types of pain receptors. Most of these receptors connect directly to the spinal cord and synapse with neurons that project from there into the brain. Some touch receptors reach densities of over two thousand per square centimeter, and there are seventeen thousand of these cells in the hand alone. Two specialized types of skin are the surfaces of the tongue and the inner lining of the nose, which contain receptors for taste and smell, respectively. The olfactory receptor neurons are by far the more numerous; over ten million of them connect the nose directly to the brain. This means that despite low average firing rates of about three spikes per second, these cells still collectively transmit more electrical impulses to the brain than either the eyes or the ears do.

We see that the volume of sensory information converging on the brain comprises tens of millions of action potentials every second, reflecting the permanent and powerful connection our brains have to our surroundings. To appreciate the magnitude of this quantity, consider again that the input from a single eye to the brain is comparable to the data transmitted over an active internet connection. If that is so, then the contributions from all our senses combined is probably worth more than ten standard internet connections, bringing in around ten megabytes of data per second over millions of neural fibers. This much data directed at a typical home computer system today could easily be enough to saturate it; hackers sometimes use exactly this approach to target and overwhelm internet sites in what is known as a denial of service attack. By analogy, our sensory environments seem to be waging a continuous denial of service attack on our brains.

Interestingly, as assessed in terms of the sheer number of action potentials, the amount of sensory input coming into the brain is also comparable to the brain's total output—the constant signaling that passes down from the brain to the rest of the body, initiating movement and regulating muscle tone. Most of the brain's motor output is relayed by the so-called *pyramidal tracts*, which consist of over one million axons firing at average frequencies of around ten to twenty spikes per second, again resulting in tens of millions of spikes per second in aggregate. From an outside observer's perspective, the brain could be seen as a somewhat convoluted mechanism for converting tens of millions of input signals per second into a roughly equal number of output signals, like a television set that converts input from your cable or antenna into moving pictures you can watch.

How do all of the incoming spikes affect the brain itself? Because the brain has evolved to accept this input, the sensory blitz does not really amount to an attack. The brain is not incapacitated—but it is changed. For sensory systems other than smell, the brain's port of entry is a structure called the *thalamus*; olfactory signals proceed through a region called the *olfactory bulb* instead. These areas in turn connect to areas of the cerebral cortex such as the primary visual cortex (abbreviated V1), in the brain's occipital lobe, or the primary auditory cortex, in the temporal lobe (see Figure 7). The influence of sensory input is felt well beyond these parts, however. More than 40 percent of the cortex is thought to be dedicated to sensory processing. In the visual system, which is the most extensive sensory modality in humans, information spreads from VI into two sets of brain regions that each pick out different features of every stimulus. In the so-called dorsal stream, areas running along the top of the occipital and parietal lobes distinguish visual stimuli on the basis of gross properties such as their location or motion in space; the ventral stream, which runs along the base of the brain's occipital and temporal lobes, specializes in more fine-grained analysis, such as the recognition of specific objects or faces. Similar streams of hierarchical processing regions handle sounds, smells, tastes, and touch.

Like gossip spreading among far-flung friends or family, the brain's incoming sensory signals eventually reach almost everywhere. Even the sophisticated high-order sensory regions maintain responses to extremely simple stimuli. Images of flashing lines, for instance, produce neural responses throughout the entire dorsal and ventral visual processing streams. More surprisingly, like the medieval barbers who also pulled teeth, brain regions that specialize in processing one type of stimulus can actually respond to other stimuli as well. For example, researchers have shown that neural signals in the visual cortex can also indicate auditory stimuli; other studies have shown responses to both visual and tactile stimuli in the auditory cortex. Brain areas known for nonsensory roles also respond to simple sensory stimuli. Parts of the frontal cortex, seat of the brain's "executive functions," are among areas activated by basic visual or auditory input. Visual responses in frontal regions can be observed even under anesthesia, showing that sensory stimuli can reach far into our brains even if we are not aware of them. A remarkable phenomenon first noted by neuroscientist Mark Raichle in the early 2000s is that a constellation of disparate brain regions is also consistently *deactivated* by many stimuli—in other words, sensory input appears to decrease the level of neural activity in these regions. The deactivated areas cover a sizeable fraction of the cerebral cortex and consist mostly of territory outside the commonly recognized sensory or motor processing systems. These regions have been termed the brain's *default mode network* because they appear to be most active when nothing noteworthy is going on.

Even subtle environmental influences can significantly affect what goes on in the brain. Most neurobiological studies of sensory responses are anything but subtle; they are performed using strong, short-lived stimuli designed to evoke brief episodes of brain activation. For instance, a visual response might be examined by measuring neural spiking or fMRI changes while a computer screen alternates every couple of seconds between a boring gray field and a bright red-and-green checkerboard. To study brain dynamics without such obtrusive stimuli, researchers perform a very different kind of experiment, in which they monitor a subject's brain under constant conditions, over many minutes, while he or she is simply lying passively in a scanner (and encouraged not to fall asleep!). The resulting *resting state* data generally display very slight fluctuations in the neuroimaging signals at each point in the brain. If you watch a crowd at a sporting event, you can probably figure out which groups of people are rooting for the same team based on who cheers or boos together, even if you have no idea what is going on in the game itself. Using similar logic, researchers try to discern which brain regions "go together" by looking for pixels that increase or decrease in intensity at the same time. Such correlations are thought to reflect neural activity in distinct brain networks and are referred to as resting state *functional connectivity*.

Resting state imaging studies show that changes in brain activity occur during continuous visual stimulation of the sort that impinges on us all the time. Neurologist Maurizio Corbetta and his colleagues collected fMRI and MEG data from subjects who passively watched either a blank video screen or clips from a spaghetti western meant to simulate naturalistic sensory experiences. Analysis showed that functional connectivity patterns differed significantly between the two conditions. During the movie segments, the correlations among neuroimaging signals were reduced across multiple brain-wide networks, revealing how ongoing sensory experiences perturb dynamics over disparate brain areas. Importantly, stimuli much less exciting than Clint Eastwood films also alter brain activity. Tamara Vanderwal of Yale Medical School presented subjects with a video featuring abstract, continuously changing shapes—images you might see on a computer screen saver—and tested its influence on resting state fMRI fluctuations. Despite their thoroughly bland content, Vanderwal's videos nevertheless perturbed functional connectivity in brain networks associated with vision, attention, and motor control. Another study, led by researchers at MIT, showed that meaningless acoustic noise can also affect functional connectivity patterns. The background noise affected regions in Raichle's default mode network, demonstrating again that seemingly trivial sensory factors can perturb gross signatures of brain function.

The science of sensory processing thus helps us appreciate why the three wise monkeys might be worried about what comes in through their eyes and ears. The sensory organs conduct a ceaseless torrent of millions of neural impulses into our brains each second. The brain has no barrier against this flood of environmental input. Even the most banal sensory signals seem to worm their way into the most exclusive tiers of the cerebral cortex—regions such as the frontal lobe, which evolved in comparatively recent history and distinguishes both monkeys and humans from most other mammals. This does not yet prove, however, that signals from the environment exert control over us. When we see sensory influences permeating the deepest reaches of the brain, we may simply be witnessing how information reaches the proverbial homunculus. Perhaps a homunculus-like neural mechanism in each of our brains preserves our autonomy even in the face of the sensory onslaught. To address this possibility, we must examine the extent to which our behavior itself is determined by stimuli from the surroundings.

The writer Albert Camus saw that people are more slaves than masters of their environment. In his novel *The Stranger*, Mother Nature drives Camus's antihero Meursault so far as to commit murder. In the book's climactic scene, Meursault encounters an adversary on a beach in Algiers and shoots him dead. "All I could feel were the cymbals of sunlight crashing on my forehead and, indistinctly, the dazzling spear flying up from the knife in front of me," the protagonist narrates. "The scorching blade slashed at my eyelashes and stabbed at my stinging eyes. That's when everything began to reel. The sea carried up a thick, fiery breath. It seemed to me as if the sky split open from one end to the other to rain down fire. My whole being tensed and I squeezed my hand around the revolver." Meursault comes to the fight armed with a gun and primed for violence by earlier events in the book. But at the moment of truth, it is the blazing sun and roaring waves that stimulate him to pull the trigger, rather than hot blood or malicious forethought originating in his brain. Through Camus's prose, we witness a man whose mind is not his own. "Meursault's crime seems utterly will-less, undetermined, and unfree," writes Camus scholar Matthew Bowker.

One might say that the heat made Meursault do it. In this respect, the fictional Frenchman in Algiers has something in common with real-life police officers in Amsterdam. In a 1994 study, a group of psychologists monitored Dutch police as they performed a training exercise in a room with variable temperature. The officers demonstrated increasing signs of hostility and belligerence when the room got hotter. Most strikingly, they showed a 50 percent greater tendency to shoot at mock offenders when the temperature was 81 degrees Fahrenheit than when it was 70 degrees. This was not a fluke finding. In an ambitious survey of sixty different studies linking climate to various forms of conflict, researchers led by Solomon Hsiang at Princeton presented evidence that higher temperature leads to increased hostility and violence across a huge range of geographic settings and time scales. Ten cases revealed correlations between temperature and violent crime or domestic disputes in particular. In one instance, the number of assaults in the city of Minneapolis was found to vary on an *hourly* basis with ambient temperature, even when time of day and other potential biases were taken into account. This points toward a biological basis for some of the temperature dependence, rather than to societal factors like the density of people on the street or fluctuations in the local economy. Backing this up, brain imaging studies have shown temperature-dependent changes in functional connectivity that could relate to effects on behavior. "Although the physiological mechanism linking temperature to aggression remains unknown," wrote Hsiang and his coauthors, "the causal association appears robust across a variety of contexts."

The relationship between temperature and aggression has two features that illustrate the directness with which the sensory environment can govern behavior. First, it is a relationship few of us are consciously aware of and that we certainly do not have a conscious say in. This places the effect of temperature on aggression firmly outside our control, and limits the extent to which cognitive processes can intervene. Second, unlike responses to many artificial environmental stimuli, such as traffic lights or television programs, the behavioral effects of temperature cannot easily be explained by learning. Even untrained laboratory mice show increased aggression as a function of temperature between about 65 and 90 degrees Fahrenheit, as indicated by their tendency to bite each other when caged together in thermostat-regulated rooms. This suggests that temperature-dependent aggression is more or less hardwired, and it emphasizes our lack of freedom to acquire or shed this environmental sensitivity. One can imagine a chain of events in which temperature changes drive receptors in our skin to stimulate alterations in brain activity and neurochemistry, resulting in increased probability of hostile or violent actions, all without any semblance of control by us or our brains.

Another environmental factor that is hardwired to human behavior is light. Appreciation of this phenomenon owes much to the work of a psychiatrist named Norman Rosenthal. Rosenthal moved to the United States from South Africa in 1976 to continue his medical studies. Migrating from the agreeable climate of Johannesburg, he found the more rugged weather of New York hard to adjust to; the long nights of winter were a particular drag. Each time the season hit, Rosenthal felt his energy sapped and his productivity declining. "My wife had it much worse than me," he recalled. "She was virtually bed-ridden for some of that time." Unlike other victims of the "winter blues," however, Rosenthal's medical background put him in a position to do something about it. He became interested in circadian rhythms, the cycles of fatigue, hunger, and other biological processes that wax and wane each day, and he joined a research group studying them at the US National Institutes of Health. One day Rosenthal and his colleagues came across a manic-depressive patient who had carefully documented his mood swings and seemed convinced that they were connected to seasonal fluctuations in the amount of light available each day. "Let's give him more light," suggested Rosenthal's colleague Alfred Lewy. Sure enough, supplementing the patient's daylight hours with exposure to bright fluorescent lighting reversed the depressions he experienced during the winter months. This finding was repeated in several larger studies and led to the identification of a medical condition called *seasonal affective disorder* (SAD), a pathology which affects millions of people worldwide and is often treated using light therapy.

Ambient light levels control both mood and circadian rhythms using a dedicated visual sensory path that again evades the potential for conscious control. A special set of retinal ganglion cells in the eye responds directly to blue light, partially bypassing the normal pathway for light sensation via rod and cone photoreceptors. The special ganglion cells are connected to a brain region called the *suprachiasmatic nucleus* (SCN), so named because it sits over the spot where the optic nerves from the left and right eyes cross in the shape of a Greek letter chi (X), at the base of the brain. The retinal input influences genes in the SCN to turn on and off in regular fashion throughout the day. This genetic circadian rhythm influences neural signaling from the SCN to another brain structure called the *pineal gland*. When darkness falls, the SCN spurs the pineal to release melatonin into the bloodstream. Melatonin is a hormone that acts broadly on many of the body's physiological systems—among other things, promoting sleep. There are competing theories about how this process relates to depression during periods of low light. In one view, too much melatonin itself is a downer, and in another, it is the timing of melatonin release that becomes problematic when days are too short. Winter months marked by excess melatonin also correspond to decreased levels of a closely related neurochemical called *serotonin*. Low serotonin levels are themselves associated with depression, and antidepressant drugs like Prozac and Celexa work specifically by boosting serotonin levels in the brain.

Along with heat and light, the colors in our environment can also act through our sensory systems to influence our behavior. The painter Kandinsky once declared, "Colour is a means of exercising direct influence upon the soul," and there have been centuries of debate about his claim. Biological studies of the effects of color are sometimes traced back to a Civil War general named Augustus Pleasonton, who introduced a pseudoscientific medical technique called *chromotherapy*. Pleasonton's method centered on the idea that sky-blue light promotes healing. Chromotherapy is now a component of New Age practices but is discounted by most doctors. In psychology the influence of colors is better substantiated than in medicine, however. A notable example is the phenomenon of Baker-Miller pink, a particular shade of light magenta that appears to soothe the savage beast in people. The calming effect was discovered by Alexander Schauss in the 1960s. Schauss showed that exposure to the pink hue caused a reduction in heart rate and breathing following exercise. He convinced a local prison to paint its cells with the color, and amazingly the jail reported a subsequent sharp drop in hostility among the inmates; the prison directors Gene Baker and Ron Miller became the color's namesakes. It is difficult to rule out the contribution of cultural biases to the effects of Baker-Miller pink. The strong association of pink with femininity in American society could influence its reception in the United States, for instance. The fact that further experiments with the color have produced conflicting results supports the notion that different populations may react disparately to the stimulus.

In a rigorous study of the effects of colors on mental function, UCLA psychologist Albert Mehrabian and his student Patricia Valdez showed seventy-six different color samples of varying hue, brightness, and saturation to 250 undergraduates and asked the subjects to report their emotional responses. They found strong effects of color saturation on arousal—more saturated colors, particularly in the blue-green-yellow range, were considered to be the most arousing. In addition, specific colors differed greatly in how pleasing they were found to be; participants rated blue to purple hues significantly higher than yellows or greens. These relatively generic findings have been extended to demonstrations of the effects of colors on a variety of cognitive tasks. In one example, researchers at the Universities of Munich and Rochester showed that subjects performed significantly more poorly on IQ tests labeled with red pen than on tests marked with green or shades of gray. Although the color stimuli were too subtle for subjects to remember them, scalp electrode recordings showed that slight changes in brain activity correlated with the color-dependent behavioral effects. A 2009 paper published in *Science* magazine by researchers at the University of British Columbia provided a possible explanation for this result by showing that the color red tends to induce subconscious avoidance behavior. The experiments also showed conversely that the color blue tended to attract participants in both word game and product choice tests, and that blue stimuli also enhanced performance in a creativity task. Although these results certainly do not imply anything medicinal about the benefits of blue sky, they are oddly resonant with the beliefs of Pleasanton and the chromotherapists.

So people are like plants, it seems, blossoming or fading with the weather—or in some cases at the whim of interior decorators. Brighter days lead to brighter moods, hotter days lead to hotter tempers, and clearer days may foster clearer thinking. The environmental influences we have considered certainly act through our brains, but they are not governed by our brains. Placed in an ambience of fluctuating conditions, our brains absorb and reflect their surroundings, channeling external influences seamlessly into changing emotional states and behaviors. Environmentally driven periods of vigor and listlessness can have deep consequences in our lives. We pursue different goals when we are aroused, equanimous, ebullient, or depressed. Others then see us differently, affecting our careers and our relationships, and perhaps even influencing the fate of our genes. On an instantaneous basis, our emotional state helps determine how we react to specific contexts or stimuli, setting the tone for momentous decisions made in the blink of an eye—like whether to accept a marriage proposal, take a job offer, or jump off a bridge. At the same time, the environmental factors that mold our emotional makeup act on a scale that is slow compared with much of human thought and action. They also make use of only a tiny fraction of the sensory input our brains receive. How does the remaining input affect us?

The bulk of the sensory bombardment alters our behavior on a much shorter time scale than ambient light and heat. As with environmental factors that affect our emotions, fast-acting stimuli carry a force of their own—they are not merely dispatches to be received by our cognitive central processor. One way to see this is to examine how stimuli interfere with each other in ways that no central controller could possible invite. If you are like me, you experience this whenever you try to concentrate on work in the presence of background noise. The resulting conflict is a well-studied phenomenon because of its importance in educational settings. Psychology researchers have defined something called the *irrelevant sound effect*, which describes the disruption of short-term visual memory (critical to reading) in the presence of background acoustic stimuli. In one example of the effect, psychologist Emily Elliott administered memory tests to cohorts of different age groups while forcing them to listen to a set of irrelevant words played over and over again. She found that adult subjects performed about 10 percent worse in the presence of the irrelevant speech than in silence, while second graders performed close to 40 percent worse. Nonspeech background noises such as tones and music also compromise cognitive task performance.

Our perceptual abilities depend on interactions between stimuli, as if different inputs fight for dominance over our brains. If you are a concertgoer, you might have noticed that some people seem to appreciate music best with their eyes closed. You have probably also noticed that people almost always close their eyes when kissing. These phenomena may be due in part to the fact that shutting off visual input can heighten other perceptual abilities. Researchers at the University of London asked a group of subjects to remember letters from a visual display while at the same time testing their ability to detect a touch stimulus. The subjects detected the touch much better when the visual task was easier, showing that the visual task interfered with detection of the tactile input. In another study, scientists at the University of Jena in Germany showed that closing the eyes in either light or darkness helps improve sensitivity to tactile stimuli. One of the most bizarre illustrations of the influence of visual input over other types of stimulation is the so-called McGurk effect, an amazing phenomenon in which observed movements of a speaker's mouth seem to override perception of speech sounds. You can listen to someone imitating a sheep—*bah, bah, bah*—but when this soundtrack is paired with a video of the speaker mouthing *fa, fa, fa,* then it is the latter sound you perceive, with an *F* rather than a *B*. As soon as you close your eyes, the sound goes back to the bleating *bah, bah, bah*. In the multisensory environment, what you see is what you get, even if you are supposed to be hearing something different.

The controlling influence of the environment is thrown into sharp relief through the phenomenon of *attention*. Attention is a metaphorical spotlight that shines on the thing or things that interest us at a given moment, and it is one of the most critical cognitive capabilities we have. The attentional spotlight determines which stimuli we most readily process, remember, and respond to. Where the light shines *from* is a controversial matter, however. We refer to our attention as being given and paid, but also attracted and caught, reflecting a dichotomy between forms of attention in which we ourselves are more active than passive participants. Neuroscientists convey this distinction by speaking of *top-down* mechanisms, apparently controlled by each individual, as opposed to *bottom-up* mechanisms, guided by the stimuli themselves. The great William James wrote in 1890 that "the turnings of our attention form the nucleus of our inner self," and that the act of "voluntarily bringing back a wandering attention . . . is the very root of judgment, character, and will." James even went so far as to declare that "volition is nothing but attention," arguing that to do anything autonomously or freely in practice boils down to paying attention to it.

Bottom-up attention by nature puts the environment rather than our own brains in the driver's seat. Walking through any metropolis, we are constantly at the beck and call of external stimuli. Our heads turn instinctively toward the honk of a horn or the nearby screech of tires on asphalt. Even a distant siren places us involuntarily on alert. The smells snaking out of a local pizzeria or Chinese restaurant also call to us, perhaps clawing at our stomachs and activating hunger for a meal. At night, the winking neon signs we pass and the flash of headlights seem to draw our gaze as if they were magnets for our eyes. These are survival reflexes, and there are good reasons for them to be deterministically written into our brains. One can imagine that the same brain mechanisms were at work when the primordial human dodged a pouncing lion or tumbling boulder.

In bottom-up attention, the biology that links sensory input to behavioral responses rests not only on the sensory systems themselves but also on the brain pathways involved in flagging stimuli as important or *salient*. These pathways automatically give salient stimuli more influence, while others pass unnoticed. The most salient stimuli are often those that suggest particular benefits or threats; this explains why our attention is more easily captured by the scent of pizza than the smell of car exhaust, or by a sudden crash from a construction site rather than the slow rumble of a train. Some apparently trivial sensory stimuli gain their force by association with a salient stimulus. Ivan Pavlov's dogs learned to associate the ring of a bell with the impending arrival of food—the bell became salient, and the hounds salivated whenever they heard it. Many neuroscientists believe that salience is signaled in the brain by pulses of particular neurotransmitters. The best-studied effects arise from rewarding stimuli such as food and sex, which induce dopamine release in regions of the brain involved in motor function. Alarming stimuli by contrast seem to cause release of the neurotransmitter norepinephrine. The involvement of distinct neurochemical processes in bottom-up attentional mechanisms emphasizes again the extent that evolution has virtually hardwired behavioral responses to large classes of environmental stimuli.

In contrast to the involuntary responses evoked by salient stimuli, top-down attention is engaged by goals you set for yourself, placing it at least nominally under internal control. But even in this context, the environment you operate in plays a strong part in your behavior, influencing what you do even on a short-term basis. How this happens is brought colorfully to life in Martin Handford's *Where's Waldo?* Each page features a tiny drawing of an elflike character named Waldo, wearing a red-striped shirt and stocking cap, embedded in a huge mob of other brightly colored individuals of about the same size and shape. Our task as readers is to find Waldo, and it is not easy. Research by attention expert Robert Desimone and his colleagues shows that our visual system processes each picture in two ways. It takes in the whole scene, searching for preferred features—a blotch of red, the peak of a hat—that invite a closer look. At the same time, it zooms in on small regions at the center of the visual field, examining them to see if they directly match the impish fellow. In Handford's complicated scenes, a thousand features beckon to our glance at once, causing our eyes to dart reflexively around the page. Our eye motions are surprisingly systematic, however; both the time we spend at each spot our eyes land on and the length of jumps our eyes make between spots in the picture are predictable, driven by characteristics of the visual scene rather than any internal decision making. A similar thing happens when we look at faces. Our gaze jumps around in a pattern that systematically takes in the eyes, nose, and mouth of the subject. Again, the details of the stimulus confronting us largely determine our brain activity and behavior.

Even at a coarser level, top-down attention is not all that top-down. It is obvious, for a start, that instructions about what to pay attention to are often derived from the environment. In laboratory experiments, top-down attention is guided by explicit visual cues like flashing signs or arrows, or by instructions delivered by researchers. In ordinary, unscripted life, shifts in top-down attention and the sometimes complex sequences of actions that follow are also instigated by trivial figments of the world around us. The taste of a cookie crumb sends Marcel Proust's narrator into a seven-volume rhapsody about life, the universe, and everything, while a visit to a once-familiar old house inspires Evelyn Waugh's Charles Ryder to find religion in *Brideshead Revisited*. The ancient historian Suetonius reported that the music of a mysterious piper drew Caesar's troops to the banks of the Rubicon in 49 BCE, and that the minstrel's blast upon a trumpet then triggered the general's fateful determination to cross into Rome, beginning the civil war that led to Caesar's dictatorship and the eventual end of the Roman Republic. Deep decisions are prompted by shallow sensory experiences.

Top-down attention is highly susceptible to the encroachment of such influences. To the dismay of classroom instructors worldwide, studies of attention spans show that most people can only pay attention to a given activity for a matter of minutes before distractions win out. Neuroscientist John Medina defines a "ten minute rule" based on the principle that audiences stop paying attention to lectures after about ten minutes. To keep people engaged, Medina proposes offering emotionally salient anecdotes or stimuli at regular intervals, in effect exploiting bottom-up attentional mechanisms to discipline listeners who are supposed to be doing better on their own. When digital devices are brought into the picture, reported attention spans become even more pathetic. A widely reported study of online habits published by Microsoft in 2015 claimed that average attention spans are as short as eight seconds, largely as the result of the prevalence of all of the electronic distractions in our twenty-first-century milieu. No matter how long it takes us to lose interest in the thing we're paying attention to at a given moment, there is always another digital stimulus there for us to—wait, let me just read this message! It is no accident that some of the activities people seem to find most engrossing—playing video games, watching TV, surfing the web—are constantly in flux, throwing new stimuli at us all the time. When faced with such barrages, our attention is held not because our brains want it to be, but because the external world is so effective at keeping our brains in its sway.

A mong the stimuli we humans receive from the world, those we get from each other pack a particular punch. We are all aware of powerful influences from other people, but the extent to which these influences degrade our own capacity for self-determination is dramatic. In 1951, a young psychologist named Solomon Asch conducted a classic experiment that demonstrates the potent effects of stimuli from peers. Asch arranged subjects into groups and asked them to take a perceptual judgment task. They were shown a card with a line drawn on it. Then they were given a second card depicting three lines, and asked to vote publicly on which line matched the line on the first card. Unbeknownst to each subject, the other members of the group were decoys—actors embedded in the experiment and instructed how to vote in the task. In some cases, the decoys were all told to vote incorrectly, even though the correct answer was always obvious. In these cases, the befuddled subjects were forced to choose between concrete visual data and the apparently unanimous but misguided opinion of others in the room. Astonishingly, most of the participants rejected the evidence of their eyes in favor of peer pressure at least some of the time. Some of the subjects went with the majority no matter what, and only about a quarter of the subjects stuck to their guns on a uniform basis.

"That we have found the tendency to conformity in our society so strong that reasonably intelligent and well-meaning young people are willing to call white black is a matter of concern," wrote Asch in summarizing his experiment. But the conformists in his trials may have been no more free to act independently than a single fish is to swim against the school, or a single wildebeest is to deviate when the herd bolts at the approach of a predator. Just as inanimate stimuli like tones and images affect the brain in distinctive ways, so do social stimuli like human forms and voices. Modern neuroscience has revealed a constellation of brain areas that are apparently specialized for socially important functions such as recognition of human anatomy and processing of human language, some of which we saw in Chapter 4. These areas are somewhat akin to sensory systems for more basic functions like vision and hearing. They are complemented by potent mechanisms for forming and solidifying interpersonal relationships, ensuring that the signals we receive from other people are salient elements of our environment.

At first sight, there appears to be a world of difference between the way low-level input like a loud noise affects our behavior and the way the actions of another person influence us. In the latter case, we generally consider a response to some extent; for instance, the subjects in Asch's conformity experiments considered their own level of confidence, the reliability of others in the room, and the properties of the stimuli before making a judgment. But social stimuli are not necessarily more sophisticated than loud noises. Think about how we instinctively react to a human scream, the cry of a baby, a smile on someone's face, or an expression of fear. Functional brain imaging studies have shown that viewing emotional facial expressions in particular produces immediate consequences to brain activity that partially mimic what would be expected if the viewer herself was experiencing those emotions.

Nonemotional social stimuli can also produce virtually automatic behavioral responses. A famous example is contagious yawning, which serves no known function but has been demonstrated in both humans and chimpanzees. Another example is the phenomenon of subliminal *speech priming*, in which a rapidly spoken word alters the listener's answers to subsequent questions even if the word is not consciously perceived. For instance, the word "cow" inserted imperceptibly into a stream of incomprehensible babble would make you more likely to recognize the word "cow" when later spoken more clearly. Speech priming effects show that our responses to human language can be every bit as reflexive as our involuntary responses to less animate stimuli.

Powerful evidence for the importance of social stimuli in our environment also comes from considering what happens when these stimuli are eliminated. Today, tens of thousands of Americans are subjects in a perverse experiment to measure the effects of such social deprivation: they are prisoners in solitary confinement. Prisoners in solitary confinement are kept in spartan jail cells no more than eighty square feet in size, with meals delivered through a slot in the door and at most an hour a day outside the cell for exercise. The only human contact they have is with prison guards, during delivery of meals or escort to the exercise yard. According to the advocacy group Solitary Watch, prisoners in solitary confinement report a litany of psychological difficulties, including "hypersensitivity to external stimuli, hallucinations, panic attacks, cognitive deficits, obsessive thinking, paranoia, and impulse control problems." A prisoner interviewed by journalist Shruti Ravindran described a particular form of breakdown that takes place in solitary. "All day, every day in the Box people go off. They yell, they scream, they talk to they-self. . . . At two or three in the morning, somebody starts screaming 'Aaaaaa!' . . . you shake your head sayin', 'Another one.'" A 1972 study looked at neuroelectrical signals from prisoners and found that inmates housed alone over multiple days showed a general slowing of brain waves. Backing up behavioral findings of stimulus hypersensitivity, the recordings also showed faster brain responses to flashes of light than observed in nonconfined prisoners. Clearly the brain of an isolated person is a different species from the brain in its natural, socially enriched environment.

A slippery slope leads from the momentary stimuli that affect individuals to global cultural factors that can shape entire populations. Social stimuli on a grand scale include the effects of everything from wars, famines, and mass migrations to divorce rates, education, and the internet. These factors set a stage on which we are merely players. You may have read the "xeroxlore" that heaven is where the police are British, the lovers French, the mechanics German, the chefs Italian, and it is all organized by the Swiss, while hell is where the police are German, the lovers Swiss, the mechanics French, the chefs British, and it is all organized by the Italians. While anyone with an extended set of European friends might rebel against the crude stereotypes, the possibility of culturally conditioned behavioral patterns is very real. For instance, French people turn out to be significantly more tolerant of extramarital affairs than other national groups, according to a 2013 Pew survey of thirty-nine countries. It is extremely unlikely that this means anything about the intrinsic nature of the French brain—genetically, French citizens are actually closer to the highly monogamous Swiss than to any other European nationality. Instead, cultural traits like attitudes toward marriage are complex social stimuli that influence our brains via the environments we live in.

Neuroscientist Michael Gazzaniga argues that "the space between brains that are interacting with each other" may be part of what "holds the answer to our quest for understanding mind/brain relationships." Properties such as our perceived autonomy and free will, which some have struggled to associate with the activity of individual brains, emerge instead from what Gazzaniga describes as multilayered, distributed interactions involving many people. It is perhaps in a similar spirit that the metaphysical poet John Donne expressed his famous thesis, "No man is an island, entire of itself." In Donne's timeless words, "Any man's death diminishes me, because I am involved in mankind." If minds arise from social interactions among brains, then the alteration or death of any person's brain presents an unmistakable insult to the collective experience.

In this chapter, I have argued that brains must be considered not only in relation to each other but in relation to the entire environment. Our brains are buffeted constantly by stimuli that impinge on us from all directions. The external influences range from subtle changes in our sensory milieu to sharper impulses from both animate and inanimate sources. These influences are not merely informational dispatches to a command center in our heads; they are causative forces that slice to the deepest levels of our brains and minds. The nervous system translates environmental input into behavioral output in much the same way that the physical structure of a tree determines how sun, wind, and rain guide its growth and movements. A tree opens its leaves toward the light and sways to avoid damage from powerful gusts, but only in a twisted sense could one say that the tree controls its actions. Instead, trees and brains are both fundamentally responsive to the world around them. Neither one presents a firewall against the environment, a place where the environment's role changes from active to passive. Neither one can be understood apart from the forest of influences that envelops it.

The dichotomy between internal and environmental drivers of brain function discussed here parallels other dichotomies we have covered in Part 1 of this book: between brain-specific and body-wide aspects of mental function in Chapter 5, between localized and diffuse brain processing in Chapter 4, between complexity and tractability of neural systems in Chapter 3, and between inorganic and organic views of brain physiology in Chapter 2. In each case, the former view emphasizes ways in which the brain is different or dissociated from the rest of the natural world, leading to popular mythologies about what the brain is and how it works. Depictions of the brain as inorganic, hypercomplex, functionally self-contained, and autonomously powerful present the brain as a surrogate for the disembodied soul and feed the attitude that I have termed scientific dualism. The cerebral mystique rests on this attitude, which is pervasive even among people like myself, who believe in a material basis for the human mind. Rejecting it requires accepting the biological basis of the mind on its own terms and witnessing how the brain, body, and environment work together to shape us. Achieving this was the objective of Part 1. In Part 2, we will turn from science to society by examining why abandoning the cerebral mystique and embracing the biological mind should matter to us both as individuals and as a global civilization.

part II

THE IMPORTANCE OF BEING BIOLOGICAL

seven

INSIDERS AND OUTSIDERS

I N THIS SECOND PART OF THE BOOK, WE WILL EXPLORE
how the cerebral mystique constrains our culture by reducing problems of human behavior to problems of the brain. Traits we traditionally viewed as built-in features of our abstract minds we now attribute
to intrinsic aspects of our neurobiology. Although this picture promotes
a more scientifically informed understanding of human activity than
we previously had, it also preserves an exaggerated propensity—deeply
rooted in our history and customs—to think of ourselves as autonomous
individuals governed from within by our own brains or minds, rather
than conditioned from the outside by the environment. In this chapter,
we will examine social implications of the modern brain-centered view of
human nature and its tendency to disregard external causes of behavior.

How we perceive the interplay between internal and external forces
that shape us is a contentious topic that has consequences throughout
our culture. In political and economic circles, conservatives and liberals
are defined by the conflicting values they place on internally driven individual enterprise versus externally determined societal and environmental

variables. The resulting tug-of-war influences tax codes and welfare pro-grams, putting food on the table for some and sapping the savings of others. In the arena of criminal justice, the establishment of guilt and the prescription of punishment depend on a related contrast in attitudes about the importance of personal responsibility as opposed to the cir-cumstances that led to a crime—internal motive versus external coer-cion. While President Barack Obama stressed the role of interdependent social forces with the soaring pronouncement that "justice is living up to the common creed that says, I am my brother's keeper and my sister's keeper," his predecessor Ronald Reagan prominently enjoined Ameri-cans to "reject the idea that every time a law's broken, society is guilty rather than the lawbreaker." "It is time," Reagan declared bluntly, "to restore the American precept that each individual is accountable for his actions." Such philosophies go hand in hand with attitudes about how to encourage achievement and initiative in many other spheres of public life. Policies regarding intellectual property and government funding, for instance, reflect the tension between incentivizing individuals and foster-ing productive environments.

Conflicting opinions about the causes of human behavior also par-allel the timeless dispute over whether nature or nurture is dominant in human development. Although the argument over how we function in the here and now is somewhat separable from the issue of how we got to be this way, a view that emphasizes the individual mind or brain as an internal driver of our actions will generally place less weight on external factors like education or upbringing that could have shaped us in the past. Conversely, a view that emphasizes our sensitivity to external con-text and environment as adults will tend to place more weight on the role of nurture during childhood. These positions in turn influence parenting strategies, educational philosophies, and social priorities.

Finding objectively justified compromises between inward- and outward-looking explanations of our actions is thus a broadly significant challenge, and the key to resolving it lies in the brain. This is because the brain is an essential link in the causal chain that binds our internal biology to the environment around us; it is the great communicator that relays signals from the outside world into each person and then back out again. With-out a biologically grounded view of the brain, we overlook the reciprocal intercourse between the individual and his or her surroundings, and must instead take sides between internally and externally oriented views of human behavior. In idealizing the brain, we overplay its role as a powerful internal determinant of how people act. Conversely, by ignoring the brain, we might overstate the importance of external influences and fail to recognize individual differences. But by demystifying the brain and recognizing its continuity with the universe of influences around us, we can better appreciate how brain, body, and environment collaborate to guide our actions.

In the next sections, we will see how historically varying attitudes about the causes of human behavior have indeed led to radically different conceptions of the place of the individual in society. We will see in particular that today's brain-centered stance emerged in part as a response to a contrasting, environment-centered philosophy called *behaviorism*, which at its heyday in the mid-twentieth century sought to explain human activity almost solely in terms of parameters of the external environment. Our subsequent turn toward inward-looking brain-based explanations now skews how we think about a host of phenomena, ranging from criminality to creativity—bringing our attention back to the individual, as opposed to his or her surroundings. A more balanced perspective is one that reemphasizes the external context that provides input and context for brain function. This is the perspective that treats us as the biological beings we are, with brains organically embedded in an extended fabric of causes and effects.

The history of psychology has been a history of debate over whether human behavior should be analyzed from inside or outside the individual, and whether internal or external factors are more important influences in people's lives. Over the past 150 years, competing schools of thinkers have advocated opposing views and gained dominance over each other in phases, like swings of a great pendulum, bringing both intellectual and social consequences with them. The cyclical nature of these changes evokes ideas of the great German philosopher G. W. F. Hegel and his intellectual descendant Karl Marx, who saw history as series of dialectics, or conflicts between antithetical agencies, such as slaves and

FIGURE 11. Famous psychologists of the nineteenth and twentieth centuries: (top left) Wilhelm Wundt; (top right) William James; (bottom left) John B. Watson; (bottom right) B. F. Skinner.

masters or workers and capitalists. In the political realm, Marx predicted that the back-and-forth struggle between classes would end with the institution of perfect equality and the universal "brotherhood of man"—a vision of social harmony that certainly has not been achieved. In psychology, however, the synthesis that reconciles internally versus externally focused views of behavior may yet be possible. Reaching this requires a biologically grounded view of the brain as mediator, however, and this has been largely lacking from major strands of psychological research.

In the late nineteenth century, the study of the mind was "virtually indistinguishable from philosophy of the soul," writes historian John O'Donnell. Even as philosophers turned to experiments, and psychology became a science, the discipline remained true to its name, from the ancient Greek word *psyche*, meaning "spirit." To the so-called fathers of scientific psychology—most notably including Wilhelm Wundt in Germany and William James in the United States (see Figure 11)—the object of study was individual subjective consciousness, and the primary research method was introspection. The birth of modern psychology was thus firmly rooted in the notion of the mind as a self-contained entity to be examined from within, extending the tradition René Descartes began when he inferred his own existence from introspection with the immortal axiom "I think, therefore I am."

Wilhelm Wundt wrote the first textbook in experimental psychology and founded what is regarded as the world's first modern psychology laboratory at the University of Leipzig in 1879. In his youth, Wundt supposedly had a habit of daydreaming that interfered with his studies but may have prepared him for his later calling. He became interested in scientific research as a medical student under the guidance of his uncle, a professor of anatomy and physiology, and then went on to secure an enviable position as assistant to the revered physicist and physiologist Hermann von Helmholtz. As he grew more independent, Wundt brewed this mix of influences into a meticulous program of inquiry into the internal nature of the human mind.

Wundt's research relied on introspection to dissect the structure of consciousness into fundamental elements such as feelings and perceptions, an approach that came to be known as *structuralism*. "In psychology," Wundt wrote, "the person looks upon himself as from within and tries to explain the interrelations of those processes that this internal observation discloses." In the laboratory, Wundt and his research team conducted experiments designed to probe the mental responses evoked by carefully adjusted external stimuli. In an effort to make their psychology more like physics, Wundt's team made heavy use of precision instruments that now appear peculiar and passé. Tools of the trade included the tachistoscope, a device designed to expose a subject to visual stimuli for extremely brief periods of time; the kymograph, an automatic mechanical data recorder; and the chronoscope, an ultrafast stopwatch designed to time millisecond intervals.

In a typical experiment, Wundt's students might record the duration of a mental process by positioning a study participant carefully in front of the tachistoscope, briefly flashing a visual stimulus, and then measuring the delay before which the subject reported his own subjective perceptions. Results would then come as a list of reaction times: thirty milliseconds for recognizing a color, fifty milliseconds for recognizing a letter, eighty milliseconds for making a choice, and so on. Broader conclusions were based on simple measurements obtained from multiple subjects under varying experimental conditions. For instance, Wundt noticed that letters written in the German gothic typeface of his time took longer to recognize than letters in Roman font, but that words in the two scripts were comprehended at the same rate; from this he inferred that the cognitive operation of reading a word does not depend on recognition of its individual letters.

Although Wundt's experiments thus targeted the interface between worlds external and internal to the individual, his interest in the internal processes of the mind was paramount, and manipulating the external sensory world was merely a means to an end. External stimuli were a vehicle for setting mental processes in motion, and these were the true objects of study to him. According to Wundt, experimental psychology "strikes out first along that which leads from outside in," but it "casts its attention primarily . . . toward the psychological side."

Wundt also dismissed the need to study the brain and thus avoided delving into the processes by which stimuli meet and alter the mind of the individual. He associated the brain more closely with the external physical domain, connected to the universe of stimuli but distinct from the domain of inner psychological phenomena. Wundt felt that efforts to connect these two domains were largely speculative and unnecessary for understanding the mind. Consideration of brain physiology "gives up entirely the attempt to furnish any practical basis for the mental sciences," he wrote in 1897. Views like these were energetically proselytized and elaborated by Wundt's student Edward Titchener, an Englishman who studied in Leipzig and imported structuralism into America when he later settled at Cornell University. Titchener believed even more completely in the need to study the mind from within. "Experimental introspection is . . . our one reliable method of knowing ourselves," he confidently proclaimed; "it is the sole gateway to psychology."

William James was a contemporary of Wundt and Titchener who taught the first psychology class at Harvard University and supervised

Harvard's first doctorate in psychology. He was the scion of a wealthy and cultured New England family, but in intellectual terms a self-made man who dabbled in art, chemistry, and medicine before adopting an academic specialty that was still only partially born. Despite being brothers in arms on the same intellectual frontier, James had few kind words for the work of his noted European counterparts in the study of the mind. The hairsplitting experimental approach of Wundt and his coworkers "taxes patience to the utmost, and could hardly have arisen in a country whose natives could be bored," James complained. Instead he advocated a more practical psychology, focused on understanding the mind and brain in terms of the functions they evolved over time to perform, rather than the components they are made of.

James had no use for the gizmos and gadgets of the structuralists, but he agreed wholeheartedly with their stance that the mind should be examined from within. "Introspective Observation," James wrote, "is what we have to rely on first and foremost and always . . . I regard this belief as the most fundamental of all the postulates of Psychology." In contrast to Wundt, James pledged allegiance to a biological view of the mind by beginning his 1890 magnum opus *Principles of Psychology* with two chapters about the brain. But although James postulated that mental activities correspond to processes in the brain, he struggled to explain their causal relationship. "Natural objects and processes . . . modify the brain, but mould it to no cognition of themselves," he asserted. The exterior world could thus influence the physiology of the nervous system, but control of conscious cognition remained within. Ultimately James cast his lot in with mind-body dualism, explaining "that to posit a soul influenced in some mysterious way by the brain-states and responding to them by conscious affections of its own, seems to me the line of least logical resistance."

The inward-looking attitudes of James, Wundt, and Titchener permeated the psychological community of their time. Their ideas spread across a sea of copious writings; Wundt alone is said to have authored more than fifty thousand pages of text. The teachings of psychology's founding figures also crossed literal oceans, as many of the students trained at Harvard, Leipzig, and Cornell came to populate the first psychology departments to emerge in America and Europe. While some of the second generation departed from their mentors' dedication to introspection and the study of individual consciousness, others found ways to influence the wider world with attitudes that sprang from precisely these approaches. The public face of psychology therefore mirrored its private academic form, organized around the self-regarding mind, shaped barely if at all by its situation in society or nature.

As a result, applied psychology at the turn of the twentieth century came to be dominated by an *essentialist* view of human nature—the idea that human capabilities and dispositions are inborn and often immutable. The most famous realization of this attitude was the emerging industry of mental testing, which was based on the concept that intelligence is an innate quality that can be measured objectively. Charles Spearman, Edward Thorndike, and James Cattell—all protégés of Wundt and James—were active in developing the first intelligence tests and the methods for interpreting them. Thorndike's tests in particular became widely applied in the US military.

Several psychologists of this time were also involved in a related brand of essentialism: the eugenics movement. A notable advocate of both intelligence testing and eugenics was Harvard psychologist Robert Yerkes, who had trained with Wundt's student Hugo Münsterberg. Yerkes argued that "the art of breeding better men, imperatively demands measurement of human traits of body and mind." Cattell was another prominent eugenicist; he had once been a research assistant to Sir Francis Galton, the British scientist and polymath who coined the term *eugenics* in 1883. Less controversially from today's perspective, Cattell was also a fervent advocate for academic freedom and personal liberty, individualist causes again resonant with the psychology of the time. This activism led Cattell to campaign publicly against US conscription in World War I; in doing so, he attracted such censure that he was forced to resign his professorship at Columbia University in 1917. Even in an age where the inborn attributes of individuals were most prized, there were limits to an individual's ability to rebel against his surroundings.

The early twentieth century was rife with rebellions and revolutions. Even before the Balkan conflicts went viral and ignited global conflagration, China had liberated itself from four millennia of dynastic rule, Irish guerillas had taken up arms against Britain, and Mexico had plunged into a decade of civil conflict. The Great War itself would see the fall of monarchies and the reorganization of class structures throughout Europe and beyond. Working men and women found ways to cast off their chains. From Talinn on the Baltic to Dubrovnik on the Adriatic, an archipelago of new nations would rise from the rubble of the continental empires. Meanwhile, Ottoman West Asia was to disintegrate into a patchwork of synthetic states that simmer with discontent to this day.

Revolution arrived in the field of psychology as well in the spring of 1913, with the publication of a polemical manifesto written by a Johns Hopkins University professor named John B. Watson (see Figure 11). The manifesto was officially entitled "Psychology as the Behaviorist Views It," and it appeared as a seemingly innocuous nineteen-page article in the journal *Psychological Review*. Its first paragraph, however, unleashed a volley of iconoclastic declarations trained not only on the psychology of Wundt and James but also on the primacy of humankind itself:

> Psychology as the behaviorist views it is a purely objective experimental branch of natural science. Its theoretical goal is the prediction and control of behavior. Introspection forms no essential part of its methods, nor is the scientific value of its data dependent upon the readiness with which they lend themselves to interpretation in terms of consciousness. The behaviorist, in his efforts to get a unitary scheme of animal response, recognizes no dividing line between man and brute. The behavior of man, with all of its refinement and complexity, forms only a part of the behaviorist's total scheme of investigation.

With this salvo, Watson announced his intention to invert the priorities of his field by creating the new science of behaviorism. Watson and his followers proposed that psychologists should focus solely on studying outwardly observable behavior and its dependence on environmental factors that could be manipulated experimentally. Their goal was to establish rules by which behavioral patterns could be entrained and altered from the outside, while jettisoning all speculation about what might be going on within. Instead of dissecting the hidden spaces of the psyche, behaviorism would thus simply ignore them. Earlier work that sought to analyze human behavior in terms of unverifiable psychological categories was rejected as *mentalism*, a putdown that subsumed the subjective psychologies of the previous age. Even earlier research on animal intelligence was considered to be too colored by mentalist analysis—prone to the sorts of anthropomorphism that might fly well in folktales but that could not be tolerated in science. In Watson's new movement, humans and animals became equally subject to dispassionate behaviorist observation and manipulation.

Initial reaction to Watson's rallying cry was mixed. Many members of the psychology establishment resisted the call. Titchener regarded behaviorism as a crass effort to "exchange science for a technology" concerned more with behavioral control than with understanding anything. Robert Yerkes sharply protested Watson's efforts to "throw overboard . . . the method of self-observation," while Columbia's James Cattell—no stranger to controversy himself—accused Watson of being "too radical." But there were many others both inside and outside the field who found Watson's stance attractive. Historian Franz Samelson has theorized that behaviorist sentiments played well to rising interest in social control in the wake of World War I, and that Watson's rhetoric won converts by "combining the appeals of hardheaded science, pragmatic usefulness, and ideological liberation." And although European psychologists notably held out against behaviorism, the alternatives they offered, such as the Gestalt school of Max Wertheimer and others, proved to be less influential. Meanwhile, the patient-focused analytical psychiatry of Sigmund Freud and Carl Jung caught the public's fancy but never gained force in scientific circles. It was behaviorism that came to dominate American psychology, effectively banishing introspection, consciousness, and other internal cognitive processes from intellectual discourse for over fifty years.

If Watson was the Moses of this movement, the great Russian physiologist Ivan Pavlov was its burning bush. In their efforts to understand how the environment governs behavior, the behaviorists took inspiration from Pavlov's world-renowned research on experimental manipulation of reflexes. The essential element of Pavlov's paradigm was the stimulus-response relationship—the ability of certain environmental stimuli to elicit reproducible, virtually automatic reactions in an animal. Pavlov most notably studied the ability of food stimuli to evoke the response of salivation in dogs; this was an innate relationship that does not need to be learned. Pavlov found also, however, that new stimulus-response relationships could be artificially induced or *conditioned*. In a procedure now referred to as classical conditioning, Pavlov achieved this effect by pairing an arousing stimulus, such as the scent of meat, with a previously neutral stimulus, such as the ringing of a bell. After repeated tests in which the bell preceded the food, the dog would learn to salivate at the sound of the bell. In this way, the innocuous bell, a formerly unimportant part of the environment, came to exert control over an animal's behavior.

Watson believed that such stimulus-response relationships could explain most forms of behavior, even in people, and that conditioning could entrain activities of arbitrary complexity. According to this view, the environment was a far more powerful factor than any internal quality in determining the behavior of an individual, at least within the bounds of what a given species could be capable of. In a self-acknowledged flight of hyperbole, Watson speculated that he could take any healthy infant at random "and train him to become any type of specialist [he] might select—doctor, lawyer, artist, merchant-chief and, yes, even beggar-man and thief, regardless of his talents, penchants, tendencies, abilities, vocations, and race of his ancestors." This phrasing tacitly acknowledges the special sensitivity of infants to training and conditioning, a feature modern researchers now describe in terms of critical periods of early development, during which the brain may be easily reconfigured. Watson and other behaviorists were little inclined to delve into such neural processes, however; avoiding invasive techniques such as electrode recordings and microscopy of brain samples, they restricted their analysis to behavioral phenomena they could more readily observe and control.

Watson's career was cut short after a scandalous extramarital affair with his research assistant forced him to resign his academic position in 1920, but a second generation of scientists soon emerged to reassert behaviorist perspectives in psychology. The leading spokesperson for these neobehaviorists was Burrhus Frederic (B. F.) Skinner (see Figure 11), an academic and popularizer who has been called the most influential psychologist of the twentieth century. Skinner promoted an experimental approach called *operant conditioning*, in which animals learn to associate their actions with intrinsically rewarding or aversive stimuli. In a typical example, a rat is placed in an unfamiliar mechanized box with a lever at one end. Every time the rat presses the lever, a food pellet is automatically dropped in for the animal to eat. At first the rat presses the lever only randomly, as part of its exploration of the box, but after a few rewards the animal learns that getting the food depends on pressing the lever. The lever pressing becomes more frequent and purposeful, and the behavior is said to have been *reinforced*. Using this method, animals can be trained to perform quite complex tasks, such as running mazes and making perceptual judgments. Skinner viewed operant conditioning as a teaching technique by which all sorts of human behaviors can also be established, from riding a bicycle to learning a language.

Behaviorists like Skinner wanted results not just in the lab but also in the wider world, which they pursued by exporting training methods from their laboratories. Several behaviorists developed educational strategies based on their science. A student of Skinner's named Sidney Bijou experimented with using rewards and punishments such as the now-ubiquitous "time-out" to discipline and teach children. Bijou's methods helped seed a more general approach known as *applied behavioral analysis* (ABA), which applies conditioning methods to improve behavior in contexts ranging from eating disorders to mental diseases. Offshoots of ABA remain in use today. Behaviorist principles also led to the development of so-called teaching machines. The first automatic teaching aids had been designed in the 1920s, but Skinner himself took the lead in updating and advertising the concept. Although his own efforts to commercialize a teaching machine met with only limited success, the publishing company Grolier successfully marketed a device called the MIN/MAX, which sold a hundred thousand units within its first two years. The MIN/MAX was a plastic box equipped with typewriter-like rollers, designed to present learning material to students through a little window. Students would answer questions displayed by the machine and then be "reinforced" with instant feedback as to whether their answers were correct or not—much like some of the teaching software available today.

The behaviorists' approach of trying to influence actions by engineering the environment was carried to even greater lengths by several prominent architects and planners of the mid-twentieth century. Writing in 1923, the Swiss-French architect Le Corbusier famously described houses as "machines for living in," a behavioristic metaphor just a few years antecedent to the introduction of Skinner's boxes. Le Corbusier and other modernist architectural trailblazers such as Frank Lloyd Wright and Walter Gropius experimented widely with open building plans and communal dwellings, through which architecture could cultivate specific patterns of domestic behavior. A number of communal settlements were also directly inspired by B. F. Skinner's novel *Walden Two*, which described his conception of a behaviorist utopia. These settlements followed behaviorist policies such as a system of reinforcing labor through earned credits and a strong egalitarian ethos that ignored intrinsic differences between people.

For behaviorists, emphasis on external versus internal control of behavior went hand-in-hand with aversion to brain science. Ironically, John Watson himself had performed dissertation research on brain-behavior relationships in rats, but he later disavowed the need for special attention to the central nervous system. Instead, he advocated a more holistic biology, akin to the view I presented in Chapter 5 of this book. As Watson put it, "The behaviorist is interested in the way the whole body works" and should therefore be "vitally interested in the nervous system but only as an integral part of the whole body." Watson contrasted this attitude with that of the introspectionists, who in his words treated the brain as a "mystery box" into which whatever could not be explained in purely mental terms was put. But Watson also insisted that the technology of his day was not up to the task of analyzing brain function, ensuring that the box would remain a mystery on his watch as well. Thirty years on, Skinner offered further reasons to neglect the brain. In his view, the nervous system was simply a causal intermediary between the environment and an individual's behavior. In short, the brain is not where the action happens. As someone interested primarily in the prediction and control of behavior, Skinner argued that it is therefore both unnecessary and inefficient to study the brain. "We don't need to learn about the brain," he reportedly quipped. "We have operant conditioning."

In the behaviorist's worldview, the environment conditions the individual the way an artist paints her canvas. The environment determines the content, color, and consistency of the individual's life. While today's scientists sometimes analogize the brain to a powerful machine, the behaviorists were more likely to cast the *environment* as a machine, implementing or embodying rules of reinforcement to shape people born naively into the state of nature. The fatal flaw in behaviorism lay in this false dichotomy between the passive person and the active environment, a contrast that led to black-boxing of the individual as a substrate merely to be acted upon. Although each behaviorist experiment was based on careful attention to environmental factors, there was little consideration for the individual's part in interpreting the universe, including the role of the brain. The behaviorists also had nothing to say about purely internal processes that do not result in observable actions. There was no place for an inner life of thoughts and perceptions, just as there was no place for the brain. This was behaviorism's flavor of dualism: a separation of internal and external spaces that placed agency squarely outside the individual, inverting the dualism of predecessors like James, for whom control came from a soul or mind acting on the inside.

The philosopher John Searle mocked the behaviorist stance with a joke about two rigidly objective behaviorists taking stock after making love: "It was great for you, how was it for me?" says one to the other. There are no subjective feelings for these two—only observable behavior. Behaviorism's greatest failure was not in the bedroom, however. As behaviorist psychologists sought increasingly to explain high-level human activities in terms of low-level conditioning, they ran into increasingly serious problems with their theory itself.

In 1959, the behaviorist framework received a mortal wound. In that year, a young linguist named Noam Chomsky published one of the most spectacular takedowns in the history of science, aimed straight at B. F. Skinner. Chomsky's critique was ostensibly a review of Skinner's book *Verbal Behavior*, which attempted to explain human language in terms of operant conditioning. Skinner had contended that verbal conversation can be explained by the kinds of stimulus-response relationships that form during conditioning. In Skinner's view, reinforcement links various utterances to complex stimuli in the real world; stimuli thus come to determine the specifics of what is said in any given context. Chomsky scornfully dismissed this notion as simplistic and vague. Rather than limiting his critique to the contents of Skinner's book, however, Chomsky examined each of behaviorism's building blocks in a broader sense, deflating the approach as a whole and not only its application to language. His review of *Verbal Behavior* became a major landmark in psychology, akin to Watson's manifesto.

Chomsky argued that outside of the highly controlled environment of the behaviorist's laboratory, the concepts of stimulus, response, and reinforcement are all so poorly defined as to be practically meaningless. Situations where multiple stimuli are present and multiple activities are being performed pose particular problems. For instance, if a crying baby girl gets a cookie after wetting her diaper while playing with toys in her crib as grandpa sings to her, what determines which of the girl's myriad actions is reinforced or which stimuli become associated with the rewarding cookie? Chomsky also noted numerous examples of activities that are performed without any apparent reinforcement at all, such as the ceaseless toiling of scholars over research projects of little interest to others, with no obvious reward in the offing. How can behaviorism explain what keeps these eggheads energized? To be fair to Skinner (and reflecting on my own experiences as an academic), there are probably still external motivations that can be invoked, but Chomsky was trying to make a more general point. He insisted that actions we perform for little or no reward can only properly be explained with reference to internal cognitive factors—in other words, variants of mentalism that behaviorists regarded as anathema. Conversely, Chomsky concluded that "the insights that have been achieved in the laboratories of the reinforcement theorist . . . can be applied to complex human behavior only in the most gross and superficial way."

Chomsky's excoriation of Skinner helped catalyze a wholesale reorientation of psychology back toward studying processes internal to individuals, an upheaval known as the *cognitive revolution*. In another swing of the great pendulum, the taboos of behaviorism were lifted, and the mind became kosher again. Several of the ideas that characterized psychology in the days before Watson underwent a resurgence following the cognitive revolution. First and foremost was the notion that the mind, rather than the environment, is the main force in people's lives. Psychologist Steven Pinker summarizes the cognitivist view as one in which "the mind is connected to the world by the sense organs, which transduce physical energy into data structures in the brain, and by motor programs, by which the brain controls the muscles." Like the commanding officer we encountered in Chapter 5, mental activity realized inside the brain guides actions based on the input it receives, exercising authority, adaptability, and autonomy.

Another old idea that resurfaced with the cognitive revolution was the concept of the mind as a parcellated apparatus, replete with numerous innate components specialized for performing different tasks. This balkanization of the mind evoked the central tenets of Wundt's structuralism but deviated sharply from the behaviorist view of individuals as "blank slates" ready to be conditioned by the external world. Chomsky himself championed the thesis that the brain contains a language organ—a neural mechanism common to all people, without which verbal communication would not be possible. Cognitive psychologists theorize that the mind and brain also contain separate modules for many other functions, such as recognizing objects, arousing emotions, storing and recalling memories, solving problems, and so on. Pinker notes the similarity of this picture to traditional Western concepts of mental and spiritual life as well. "The theory of human nature coming out of the cognitive revolution has more in common with the Judeo-Christian theory of human nature . . . than with behaviorism," he writes. "Behavior is not just emitted or elicited. . . . It comes from an internal struggle among mental modules with differing agendas and goals."

After the cognitive revolution, advances in knowledge conspired with newfound emphasis on the internal qualities of individuals to promote a convergence of psychology with neuroscience. Researchers and laypeople alike came to equate mental functions with brain processes. The advent of functional brain imaging methods in the 1980s and 1990s particularly facilitated this connection and allowed neuroscientists to test hypotheses about the modularity of mental and neural organization together. In the era of cognitive science, resurgent appreciation for the complexity of the mind harmonized perfectly with the blossoming of research into what seemed like an infinitely complex organ of the mind. An equally important convergence occurred at the border between psychology and the emerging field of computer science. It was at this interface that computational theories of mental function arose, forwarded by such figures as the perceptual psychologist David Marr. Marr famously characterized the mind as an information-processing device based on the computational conversion of inputs to outputs using algorithms implemented in physical hardware. Captivated by such descriptions, many psychologists and neuroscientists of the time began to claim that "the mind is the software of the brain," extending the computational model of mental processing into the full-blown metaphor for biological function discussed in Chapter 2.

The mystique of the brain reached its apogee during this time, and it is easy to see why. Currents of the cognitive revolution exalted the brain even as they undercut the significance of the wider world. Neuroscience became a hot topic, while behaviorism became a bad word, associated both with scientific superficiality and with the state-sponsored behavioral control of places like Stalinist Russia. With behaviorism dethroned, people came to think foremost of the brain when striving to get to the root of mental properties ranging from artistic genius to drug addiction, and it was correspondingly less common to consider extracerebral influences.

Some commentators have begun to use the word *neuroessentialism* to describe this phenomenon. "Many of us overtly or covertly believe . . . that our brains define who we are," explains philosopher Adina Roskies in her definition of the neologism. "So in investigating the brain, we investigate the self," she writes. This idea that the central nervous system constitutes our essence as individuals echoes the earlier essentialist attitudes of people like Wundt and James, who conceived of the mind as a set of inborn attributes, or of Yerkes and Cattell, who advocated the measurement and breeding of innate qualities as part of the intelligence testing and eugenics movements of the early twentieth century. The similarity between old and new views may indeed explain why modern psychology and neuroscience appear to be so compatible with traditional Western concepts of the soul.

The cerebral mystique and the scientific dualism we considered in Part I promote neuroessentialism by emphasizing soul-like qualities of the brain—its inscrutability, its power, and perhaps even its potential for immortality. The biochemical continuity and causal connections among brain, body, and environment are lost, and the brain alone comes to occupy the role of commander and controller. As a result, the divide between internal and external influences remains almost as stark as in the days of the earlier *-isms*, when the brain was largely ignored.

We have now seen how the cognitive revolution and the rise of neuroscience carried the brain to its central place as an explanatory factor in our lives. The neuroessentialist attitude that our key characteristics are determined by our brains reflects the continuing backlash against behaviorism and its emphasis on the environment. Like behaviorism, however, the modern brain-centered perspective often fails to produce an integrated picture of how internal and external factors combine to guide human activity. Instead, it encourages a focus on the role of the brain to the exclusion of other factors that influence what people think and do. We can appreciate this most clearly by considering a specific example of neuroessentialism in action.

The world's most notorious icon of neuroessentialism spent its golden years steeping placidly in a jar of formaldehyde, secluded from civilization in a brain bank at the University of Texas at Austin. The specimen's peaceful retirement belied a violent past, however; this past was stamped indelibly into its stripes of gray and white matter like stains and creases on an old piece of newspaper. For a start, the famous brain was far from intact. Soon after the organ had been extracted from its natural habitat, a pathologist's knife sliced it into fillets, unkindly exposing its inner structures for an autopsy. Many of the cuts revealed gross disfigurations. The left frontal and temporal lobes, once seats of cognitive and sensory prowess behind the eye and ear, were brutally lacerated. Shards of bone had sheared across the tissue, forced through by the impact of hot lead bullets on the once protective skull. In another part of the brain, near a pinkish blotch called the *red nucleus*, the tissue had been pried apart by another kind of bullet—a malignant tumor the size of a walnut that would have meant certain death to its bearer, had metal projectiles not arrived first. The story of this tumor and the organ that harbored it illustrates how the problem of reconciling the agencies that act inside versus outside the individual remains unresolved even now.

The forlorn brain's original owner, a former marine and trained sniper named Charles Whitman, had committed one of the worst mass murders in American history. In the early hours of August 1, 1966, he stabbed his mother and his wife to death. Later in the morning, he transported a small arsenal of weaponry to the top of the 307-foot Main Building on the UT Austin campus and unleashed a fusillade that killed or wounded forty-eight people in and around the tower. Two hours into the rampage, Whitman was finally cornered by Austin police and felled by two blasts from Officer Houston McCoy's shotgun. To the public, not accustomed to outbreaks of military-style violence in civilian settings, the massacre was deeply unsettling. "In many ways Whitman forced America to face the truth about murder and how vulnerable the public could be in a free and open society," writes author Gary Levergne in his book *A Sniper in the Tower*.

But it was Whitman's brain that soon claimed center stage. In the months before the murders, Whitman had experienced painful headaches and also began seeking help from a psychiatrist. In a suicide note, he speculated that he was suffering from a mental disorder and urged investigators to perform a thorough autopsy to determine what could be wrong with him. When postmortem examination revealed the tumor near Whitman's hypothalamus and amygdala, brain regions involved in emotional regulation, some seized on it as a possible explanation for his inexplicable act of destruction. Many of Whitman's friends and family were particularly ready to believe that the brain tumor had changed his personality for the worse—that his diseased brain had *made him* commit the heinous crime.

Amygdala expert Joseph LeDoux thinks that the mere possibility of tumor-induced behavioral effects might have been enough to reduce the murderer's sentence, if he had survived to trial. Neuroscientist David Eagleman goes much further; he argues that cases like Whitman's uproot our notion of criminal responsibility, because they demonstrate the causal role of brain biology in behavior. We do not control our biology, so how can we be held responsible for it? Eagleman predicts that soon "we will be able to detect patterns at unimaginably small levels of the microcircuitry that correlate with behavioral problems." Eagleman's speculation is not too tendentious. In one of the most talked-about applications of neuroimaging technology, it has already become possible to look at criminals' brains in situ. "Lawyers routinely order scans of convicted defendants' brains and argue that a neurological impairment prevented them from controlling themselves," writes judicial scholar Jeffrey Rosen. The movement has gone so far, he explains, that "a Florida court has held that the failure to admit neuroscience evidence during capital sentencing is grounds for a reversal."

These perspectives powerfully illustrate the influence of neuroessentialism, expressed here in the notion that there could be inherent and perhaps immutable properties of a person's brain that are enough to explain the nature of his or her acts, whether criminal or otherwise. The consequences of this conclusion are far-reaching. If our brains give rise to actions that are beyond our conscious control, then how can we continue to hold individuals accountable for what they do? Rather than judging the person guilty or innocent of his crimes, we should judge the brain guilty or innocent. Rather than punishing the perpetrator for biology beyond his control, we should remedy whatever fault we can find in his brain, or failing that, we should prescribe imprisonment merely as a means of protecting society, just as we might impound an unroadworthy car. Stanford neurobiologist Robert Sapolsky quips that "although it may seem dehumanizing to medicalize people into being broken cars, it can still be vastly more humane than moralizing them into being sinners."

Charles Whitman's brain, lazing quietly in a closet in Austin, served as a tangible symbol of these ideas—until one day it vanished. The murderer's cerebral matter had been part of a collection of around two hundred neurological specimens bequeathed to UT Austin in the late 1980s. Roughly thirty years later, photographer Adam Voorhes and journalist Alex Hannaford sought out the brain in connection with a book about the collection. Their research prompted the revelation that about half of the specimens, including Whitman's, had gone missing. "It's a mystery worthy of a hard-boiled detective novel: 100 brains missing from campus, and apparently no one really knows what happened to them," wrote Hannaford in the pages of the *Atlantic*. Amid widespread media attention, UT researchers struggled to find an explanation and soon reported that the lost brains had probably been disposed of as biological waste in 2002.

The strange disappearance of Whitman's brain inspires a thought experiment to probe the limitations of neuroessentialism: Let us ask ourselves what would have happened if the brain had been erased even more completely from the picture. What if Whitman's brain had vanished before the doctors got to it, and its malignant growth, or indeed any other neural abnormality, could not be invoked in order to rationalize his crime?

The answer is that we would be forced to look elsewhere for factors that could have contributed to the events—and we would easily find them in the records of the case. We would perceive, for instance, an atmosphere of social tension running throughout Whitman's life. We would recognize the killer's difficult relationship with his father, a martinet who beat Whitman's mother and wrecked their marriage just months before the shooting. We would witness Whitman's repeated career rejection—how he was forced to terminate his academic studies because of poor performance and later faced a humiliating court-martial and demotion in the marines. We would observe Whitman's record of substance abuse (he had a jar of amphetamine with him on the day of the shootings). We would register the ease with which Whitman procured his weaponry, and we would appreciate his immersion in violent culture. We would see that Whitman had physical stature and strength without which he would not have been able to commit his acts. We might even note the scalding 99-degree Fahrenheit temperatures recorded around the time of the murders, which could have played a part in exposing the murderer's latent aggression. In short, numerous circumstances around the killer could have contributed to his crime just as surely as his brain did.

The case of the Texas Tower murders is emblematic of the dichotomy that still persists between internal and external explanations of human behavior. There are two accounts of the perpetrator's actions, one narrated from the inside and the other from the outside. One is a subjective story that could have been told by the nineteenth-century fathers of psychology we met earlier in the chapter—just substitute "brain" for "mind," and it all falls into place. The other is a tale that sounds more like something John Watson or B. F. Skinner would have come up with. The narratives do not differ in the extent to which they assign moral responsibility to Charles Whitman. Whether the tragedy of August 1966 grew from a seed in Whitman's brain or from a seminal event in the environment around him does not affect the amount of metaphysical blame we can place on his shoulders—in either case, he is a pawn in nature's great game of life and death. The internally and externally centered accounts do differ, however, in the extent to which they focus our analysis on this one man and his internal makeup, versus the society and environment around him. They differ in the causes they lead us to consider paramount, and they differ in the ways they instruct us about how to prevent future calamities. Most importantly, they differ in the extent to which they spur us to think of justice and transgression as a matter of individuals' actions or as a matter of the interactions that affect them. Neuroessentialism focuses our attention directly on the individual and his brain, but in doing so, we miss the other half of the picture.

If we cast our net more broadly, we can find further examples of neuroessentialism in numerous walks of life. In each context, describing a phenomenon in predominantly neural terms tends to blind us to alternative narratives built from external factors that act around the brain rather than within it. When we strive for explanations centered on the brain alone, we fall for what the philosopher Mary Midgley calls "the lure of simplicity." We conceive of the brain as a colossus among causes, a plenary speaker rather than a panelist in the broader conversation among internal and external voices that might all deserve a hearing. The cerebral mystique prompts us to privilege the brain, in each situation making us less likely to consider the importance of body, environment, and society—even if we tacitly admit that they all play a role. This in turn affects how we understand and treat a range of social and behavioral problems in the real world.

Why are teenagers different from adults? We have all noticed the emotionally volatile and often reckless conduct of many adolescents, compared with adults. "Youth is hot and bold," wrote the Bard of Avon, while "Age is weak and cold; Youth is wild and Age is tame." Cognitive neuroscientist Sarah-Jayne Blakemore argues that immature brain biology may explain stereotypical adolescent traits such as heightened risk taking, poor impulse control, and self-consciousness. The teenage brain is not just a less experienced version of the adult model; evidence from neuroimaging studies shows both structural and dynamic differences between the brains of teenagers and grown-ups. "Regions within the limbic system have been found to be hypersensitive to the rewarding feeling of risk-taking in adolescents compared with adults," Blakemore explains, "and at the very same time, the prefrontal cortex . . . which stops us taking excessive risks, is still very much in development in adolescents."

Accounting for the immature behavior of teenagers in terms of the immaturity of their brains could be a risky business as well, however. For one thing, it goes without saying that a teenager's biology differs from an older person's in ways that go far beyond the nervous system. Regardless of brain disparities, hormonal and other bodily influences dramatically affect how teens feel about themselves and how they react to situations. We might also note that modern-day late teens were in the prime of life by the standards of prehistoric times, when evolution picked us apart from our apish ancestors and life expectancy was probably well under thirty. Today's late adolescent brain would have come across as comparatively grown-up back in 50,000 BCE. The most radical differences between contemporary adolescents and their late Paleolithic counterparts are cultural rather than physiological. This suggests that what makes teenagers seem immature now might also have more to do with culture than with biology. In our place and time, teenagers inhabit a vastly different world from adults. They have a quality and quantity of social interaction that few adults experience, their lives are scripted and controlled in ways few adults would embrace, and their daily goals are very different from those of their parents or grandparents. Disentangling the consequences of brain biology from these environmental factors is next to impossible. But if we are trying to understand and possibly correct the special foibles of our teenage relatives or friends, it seems simplistic to focus mainly on the idiosyncrasies of their brains.

What makes someone a drug addict? Addictive drugs are not just more potent versions of other things we enjoy, like good food or sunny days; they actually penetrate into the brain and directly change the behavior of brain cells. The past two decades have seen tremendous progress in elucidating which brain processes are involved in susceptibility to narcotics, some of which I work on in my own lab. Stressing the centrality of brain biology, the US National Institute of Drug Abuse (NIDA) now defines addiction as "a chronic, relapsing brain disease that is characterized by compulsive drug seeking and use, despite harmful consequences." Part of NIDA's objective in describing addiction in this way is to try to mitigate the moral stigma associated with substance abuse. Explaining addiction in terms of subconscious brain functions seems to relieve the addict of guilt—just as evidence of brain pathology apparently exonerates criminals like Charles Whitman.

But it is not necessary to blame brain biology in order to forgive the addict. External social and environmental variables such as peer pressure and weak family structure are well-known risk factors for addiction, as is being male or growing up poor. A person stuck in such conditions can barely be blamed more for his or her situation than a person with brain disease can be blamed for having an illness. Meanwhile, characterizing addiction as a brain disease may constrain the available avenues for treating it. Sally Satel and Scott Lilienfeld argue that the brain disease model "diverts attention from promising behavioral therapies that challenge the inevitability of relapse." Psychiatrist Lance Dodes advances a similar point, emphasizing the environmental stimuli that contribute to drug use. He writes that "addictive acts occur when precipitated by emotionally significant events . . . and they can be replaced by other emotionally meaningful actions." At an even broader level, addiction calls for solutions that involve social and cultural elements in a way that other noninfectious diseases like cancer do not. Efforts to reduce poverty, keep families together, and improve school settings might turn out to be as influential as efforts to combat addiction-related processes within the brain itself, for example using brain medicines. Addiction is a multidimensional phenomenon, and it is important to maintain sensitivity to dimensions that extend beyond the head.

What makes someone an outstanding artist, scientist, or entrepreneur? The hero of Mel Brooks's 1974 comedy *Young Frankenstein* believes that his man-made monster can become a genius by receiving the transplanted brain of the great German "scientist and saint" Hans Delbrück. But is having a great brain really what it takes to become a great contributor to society? We saw in Chapter 1 that researchers have struggled for over a hundred years to relate extraordinary personal achievements to characteristics of the brain. Science writer Brian Burrell tells us that "none of the studies of . . . 'elite' brains have been able to conclusively pinpoint the source of mental greatness," but this does not stop people from continuing the search today. Modern researchers apply neuroimaging tools to locate what psychologist and US National Medal of Science winner Nancy Andreasen calls "unique features of the creative brain." In her own work, Andreasen has discovered patterns of brain function that appear to distinguish writers, artists, and scientists from ostensibly less creative professionals. Other researchers have examined fMRI-based correlates of improvisation, innovative thinking, and additional hallmarks of creativity.

While few scientists would dispute the notion that brain biology underlies differences in many cognitive abilities, we also know that culture, education, and economic status contribute enormously to the expression of such abilities in creative acts. According to studies of creativity in identical twins, evidence for a genetic role—which would determine inborn aspects of brain structure—is ambiguous at best. Meanwhile, a country with one of the most ethnically and presumably neurally heterogeneous populations in the world (the United States) is also one with by far the largest number of Nobel laureates. This kind of statistic casts doubt on the notion that a single brain type or set of characteristics is what fosters creativity in multiple walks of life. In other words, there can be no such thing as a "creative brain."

Psychologist Kevin Dunbar has studied creativity in molecular biology labs and found that new ideas were most likely to come from group discussions involving diverse input, as opposed to from individual scientists working with their brains in relative isolation. In some cases, the principle that creativity comes from the convergence of disparate ideas creates entire fields that seem innovative, like nanotechnology or climate science. Even when individuals are working on their own, the insights they have may be prompted by exposure to a diversity of environmental stimuli. "A change in perspective, in physical location, . . . forces us to reconsider the world, to look at things from a different angle," writes journalist Maria Konnikova, who has made a study of creative processes. She explains that sometimes "that change in perspective can be the spark that makes a difficult decision manageable, or that engenders creativity where none existed before." In contrast, the conception that creative acts arise from a "creative brain" reduces the trove of biological and environmental influences down to a neuroessentialist nugget. If we are trying to understand or encourage creativity in our society, attending to the world around the brain may be just as important as cultivating the brain itself.

Where does morality come from, and what makes people perceive actions as right or wrong? One of the most fascinating manifestations of neuroessentialism lies in the resurgent movement to describe morality in terms of innate mechanisms in the brain. In 1819 Franz Gall had placed an organ of "moral sense" above the forehead, near the meeting of the brain's left and right frontal lobes. Reflecting a more recent view, Leo Pascual, Paulo Rodrigues, and David Gallardo-Pujol of the University of Barcelona explain that "morality is a set of complex emotional and cognitive processes that is reflected across many brain domains." Functional neuroimaging studies point to relationships between moral reasoning and activation of areas associated with a potpourri of processes, from empathy and emotion to memory and decision making. Such findings jibe with our intuitions about the complexity that underlies many moral problems, and they also give physical homes to the various considerations we tend to weigh.

But framing moral processing foremost in neural terms serves once again to distract from the importance of environmental and social influences. Like other choices we make, ethical decisions are highly dependent on intangible external factors, as well as on bodily states. Extracerebral factors that interact with our emotions can affect our moral calculus— for instance, by biasing us toward or against aggressive behavior. Our internal moral compass is even more dramatically swayed by social cues. Most obviously, we tend to let our guard down when we feel that nobody is looking. Conversely, we may be more likely to perform questionable actions when they seem to be socially accepted. This was stunningly demonstrated by Yale University's Stanley Milgram, who showed in a 1963 study that two-thirds of a randomly recruited group of forty male subjects were willing to administer painful 450-volt electric shocks to strangers when encouraged to do so by an experimenter in a laboratory setting. Psychologist Joshua Greene, who leads Harvard's Moral Cognition Lab, notes that the neural mechanisms involved in moral choices are "not specific to morality at all." In fact, what places a choice in the *moral* domain at all is precisely its dependence on external context and on the culturally conditioned judgments of others about what constitutes right and wrong behavior. We risk losing sight of this if we boil moral reasoning down to brain processes independent of the external features that feed into them.

The question of what makes individuals the way they are is analogous to a central issue addressed by historians: What made events unfold the way they did? The Scottish historian Thomas Carlyle gave a stark answer to this question, writing that "the history of what man has accomplished in this world, is at bottom the History of the Great Men who have worked here." Gazing across the cultural landscape that surrounded him in 1840, it seemed to Carlyle that "all things that we see standing accomplished in the world are properly the outer material result, the practical realization and embodiment, of Thoughts that dwelt in the Great Men sent into the world." This was the Great Man theory of history—the hypothesis that the minds of a few notable people transformed civilizations around them and determined the course of events. In Carlyle's metaphor, the likes of the poets Dante and Shakespeare, the scholars Johnson and Rousseau, and the tyrants Cromwell and Napoleon were nothing less than beacons of light, "shining by the gift of Heaven," and bestowing "native original insight, of manhood and heroic nobleness," on the dim multitudes around them.

In a culture now swayed by the cerebral mystique, light shines not from the Great Men of the past but from our brains. It is a picture that William James himself evoked when he wrote that all of humankind's inventions "were flashes of genius in an individual head, of which the outer environment showed no sign." A line connects James's words to the luminescent brain images in today's pop neuroscience articles, and to the reduction of problems of the modern world to problems of the brain. But rebutting the neuroessentialist thesis that the brain is an autonomous, internal engine of our actions is an equally extreme antithesis: the behaviorist's view of human endeavors as explained primarily by the environment.

Today we can bring the sparring sides together. With the cognitive rebellion against behaviorism receding into the past, and with growing understanding of the ways in which the brain interacts with its surroundings, we need no longer see the internally and externally weighted views of human nature as necessarily opposed. In the age of neuroscience, we can doubt neither the life of our minds nor the central role of our brains in it. But at the same time, we cannot doubt that external forces extend their fingers into the remotest regions of our brains, feeding our thoughts with a continuous influx of sensory input from which it is impossible to hide. We also cannot deny that each of our acts is guided by the minute contours of our surroundings, from the shapes of the door handles we use to the social structures we participate in. Science teaches us that the nervous system is completely integrated into these surroundings, composed of the same substances and subject to the same laws of cause and effect that reign at large—and that our biology-based minds are the products of this synthesis. Our brains are not mysterious beacons, glowing with inner radiance against a dark void. Instead, they are organic prisms that refract the light of the universe back out into itself. It is in the biological milieu of the brain that the inward-looking world of Wundt and today's neuroessentialists melts without boundary into the extroverted world of Watson and Skinner. They are one and the same.

eight

BEYOND THE BROKEN BRAIN

I F THE THINGS YOU DO ARE DUE TO YOUR BRAIN, THEN defects in your behavior must likewise spring from defects in your brain. This is the logic behind the recasting of mental illnesses as brain disorders—a change that has coincided with the growth of neuroscience and the rise of the cerebral mystique. Advocates for equating mental illness with brain dysfunction argue that doing this reduces the stigma traditionally associated with psychiatric problems. Viewing conditions like depression or schizophrenia as brain diseases mitigates a tendency to blame mentally ill people for their pathologies. We do not fault people with maladies of the liver or lung, so why would we fault people with maladies of the brain? "Schizophrenia is a disease like pneumonia," says the prominent neuroscientist Eric Kandel. "Seeing it as a brain disorder destigmatizes it immediately." There is evidence that redefining mental disorders in such biological terms makes people with these problems more likely to seek treatment, a hugely important outcome for patients and their friends and family. Admitting to yourself that you have an organic disease may be far easier than admitting to yourself that your soul is corrupted.

It was not long ago that the mentally ill were considered morally culpable for their acts of unreason and their defiance of social norms. According to the French social theorist Michel Foucault, Enlightenment-era Europe defined the madman as one who "crosses the frontiers of bourgeois order of his own accord, and alienates himself outside the sacred limits of its ethic." In his masterpiece *Madness and Civilization*, Foucault casts the insane not as disease victims but as social transgressors, whose violation of cultural expectations led to their confinement in virtual prisons in which they were subjected to extrajudicial discipline and deprivations of all kinds. Even nineteenth-century asylum reformers such as Samuel Tuke and Philippe Pinel continued to moralize their patients while offering them more humane treatment. "To encourage the influence of religious principles over the mind of the insane, is considered of great consequence, as a means of cure," wrote Tuke in 1813. According to Foucault's analysis, the benevolent asylum of Tuke's period was still "a juridical space where one is accused, judged, and condemned" and where madness remained "imprisoned in a moral world."

Defining psychiatric disorders as brain diseases may be one of the best ways to liberate them from the moral internment Foucault and others described. Similarly, biologically motivated treatment programs that act via defined brain-based mechanisms offer an incomparable advance over many earlier approaches of questionable merit and efficacy, such as the leg irons and water cures of earlier centuries. At the same time, mental illnesses in our day continue to pose a dramatic and surprisingly intractable challenge to society. According to statistics highlighted by the National Alliance on Mental Illness (NAMI), a large patient advocacy group, about one-fifth of American adults suffer from mental disorders each year, and serious mental illnesses cost the United States more than $190 billion annually in lost wages. Meanwhile, despite efforts to improve acceptance and access to therapy, over 50 percent of adults with mental illness fail to receive treatment each year. NAMI reports that depression alone is "the leading cause of disability worldwide" and that 90 percent of US suicides are linked to mental disorders. Clearly there is much work still to be done.

In this chapter I want to show you that the cerebral mystique is part of the reason why mental illnesses remain such a scourge. Idealization of the brain and emphasis on its causal role in psychiatric disorders contributes in three significant ways. First, in place of the stigma of mental illness, it introduces a new phenomenon that works against mental health patients: the stigma of having a "broken brain." Second, by focusing the attention of psychiatric patients, doctors, and researchers squarely on the brain, the equation of mental illness with brain disorders diminishes consideration of potentially effective therapies that do not physically enter the brain. Third, because problems with individual brains are necessarily problems with individual people, an overemphasis on neural underpinnings of mental illness understates the role of environmental and cultural contributions that extend beyond individuals; this makes it seem less urgent for us to discover and correct ambient factors that might increase the prevalence of mental disorders. Underlying each of these difficulties is the same neuroessentialist trend we saw in the last chapter: the tendency to reduce problems in our society to problems of the brain. In the context of mental health, neuroessentialist attitudes affect programs of research and medical practice that touch the lives of billions of people.

Among the most prevalent fixtures of American research laboratories are the rust-colored cardboard-bound notebooks in which virtually all biologists and chemists record their daily work. The moment when a researcher opens one of these books for the first time can be immensely intimidating; its thick folio of blank yellow graph paper represents a mountain of work that will soon need to be scaled. Over time, the sheets will gradually be filled with handwritten descriptions of experimental designs, details, and outcomes, usually supplemented by pasted-in printouts or photographs documenting key results. Once completed, such books are stored indefinitely, since they often constitute the only physical product of long and sometimes painful days and nights of hard labor at the bench. Whenever a student or postdoc leaves a laboratory for greener pastures, his or her notebooks are ritually handed off to the lab head. I have two shelves of these notebooks in my office, some of which I myself filled up when I became an independent postdoc in the late 1990s.

On July 23, 2012, a similar brown lab book was discovered lying in a mail room at the Anschutz Medical Campus of the University of Colorado. It was stashed in an envelope together with a wad of partially incinerated twenty-dollar bills. The experience of opening this notebook was especially chilling. Inscribed in it were the pathological scribblings of James Holmes, a former neuroscience PhD student who three days earlier had committed one of the worst shootings in American history at a movie theater in Aurora, Colorado. Instead of handing the notebook to his mentors at the university, Holmes had mailed it with the burnt money to his former psychiatrist before he committed his crime. And instead of using the notebook to record scientific progress, Holmes had documented his approach to mass murder, from sociopathic musings to labeled diagrams of the crime scene. His was a journey more painful than the most unlucky scientist's succession of failed experiments, and one that culminated in devastation for dozens of innocent people.

Almost every page of Holmes's book bore evidence of a nihilistic and destructive outlook, but unlike the Texas Tower murderer Charles Whitman, whose dissected brain we learned about in Chapter 7, Holmes had already concluded that his problems stemmed directly from his brain. He repeatedly referred to his brain and mind as damaged. Under the heading "Self Diagnosis of Broken Mind" on page 31, Holmes listed thirteen psychiatric conditions beginning with dysphoric mania and ending with restless leg syndrome. Extending the theme of mental fracture on page 36, he continued:

> I tried to fix it. I made it my sole conviction, but using something that's broken to fix itself proved insurmountable. Neuroscience seemed like the way to go, but it didn't pan out. In order to rehabilitate the broken mind my soul must be eviscerated. I could not sacrifice my soul to have a "normal" mind. Despite my biological shortcomings I have fought and fought. Always defending against predetermination and the fallibility of man.

As nonsensical as much of Holmes's ranting was, a consistent message was that he felt himself to be utterly and irredeemably flawed, in both mental and physical senses. After his arrest, Holmes continued to speak repeatedly about his broken brain with a court-appointed psychiatrist. At the trial itself, Holmes explained that the broken brain made it impossible for him to interact with other people, leading him to hate humanity.

Holmes displayed no reluctance to undergo treatment for mental illness, nor did he appear unwilling to acknowledge his psychiatric conditions, at least to himself and his doctors. Since his teen years he had been in contact with psychologists and psychiatrists, and he had seen a therapist at the University of Colorado as recently as a few weeks before the shootings. He had also tried a variety of medicines, including benzodiazepines, which are known for relieving anxiety, and selective serotonin reuptake inhibitors, which are widely prescribed as antidepressants. He knew that his defects were "biological" and indeed had taken up the study of neuroscience in part to address them. But awareness of his apparent brain disease was deeply disturbing to Holmes in a way that no respiratory illness or even cancer could possibly have been. He felt no apparent guilt or responsibility for his mental state, but the sense of physical worthlessness he experienced might have been even worse.

James Holmes was also far from alone in suffering the stigma of a broken brain. For a more balanced perspective than Holmes's, we could consult Jack Bragen, who was diagnosed with schizophrenia at age eighteen and now writes occasional columns on mental illness for a local newspaper in Berkeley, California. Bragen describes how the broken brain model of mental illness affects patients struggling with acceptance of their condition and the need to receive treatment. "In order to take medications, it is not absolutely necessary to admit that one's brain has a flaw," he writes, "but if one doesn't acknowledge such a neurological flaw, there isn't a reason for taking these drugs." Admitting that one has a flaw in one's brain is a painful step that in Bragen's experience "requires courage and self-worth as well as an ability to accept oneself unconditionally." Self-stigmatizing is an important issue for many people with mental illness. Psychiatric patients who perceive themselves as compromised by immutable neurological fate may feel helpless to improve their situation and may do less to help themselves as a result. This is the flip side of replacing the language of personal responsibility and willpower with the language of biological causation and neuroscience.

The increasing identification of mental illness with biologically grounded brain disease also seems to have done little to shake the public's stigmatization of psychiatric patients. A large international analysis of changing attitudes about mental illness in Europe and America from

1989 to 2009 found no improvement in social acceptance of patients with depression or schizophrenia, despite significant growth in awareness of neurobiological explanations. Embedded in these results are subtly competing influences. Psychologist Patrick Corrigan, one of the study's authors, agrees that biological explanations for mental illness may reduce people's willingness to admit bias against psychiatric patients. He and his collaborator Amy Watson suggest that "people are less likely to endorse . . . social avoidance toward people with mental illness after they have been educated about how mental illness is a biological disorder that people do not choose." On the other hand, the two stress that "biological explanations may also imply that people with mental illness are fundamentally different or less human." Under these circumstances, society may view psychiatric patients as more intrinsically dangerous or incapable of taking care of themselves. This only exacerbates the desire to keep people with mental illness at a distance, or to subject them to controls and possible institutionalization.

It is under the guise of protecting society from defective biology that the worst abuses of the mentally ill have been perpetrated. The twentieth century saw thousands of supposedly intellectually disabled people forcibly sterilized in Europe and America. A notorious example was that of Carrie Buck, an eighteen-year-old girl said to have had a mental age of nine, who was committed to the Virginia State Colony for Epileptics and Feebleminded after becoming pregnant out of wedlock in 1924. Buck had no family to protect her when she was ordered to have her Fallopian tubes cut after giving birth; her own mother had been committed to the same establishment years earlier and was also at the state's mercy. Although Buck appealed the decision all the way to the US Supreme Court, she ultimately lost. In upholding the decision to sterilize, Chief Justice Oliver Wendell Holmes wrote, "It is better for all the world, if instead of waiting to execute degenerate offspring for crime, or to let them starve for their imbecility, society can prevent those who are manifestly unfit from continuing their kind. . . . Three generations of imbeciles are enough." Even this jurist, who represented the height of progressivism in his day, thus viewed mental illness as a physical blight severe enough to deserve a remedy we consider inhuman today.

An unusual burial ceremony that took place at the German city of Tübingen in July 1990 bore witness to the most extreme consequence of stigmatizing brain disease. At this funeral, the deceased was not an individual but rather a set of scientific specimens, including brain sections obtained from victims of the Nazi program of "euthanasia" of the mentally disabled. These specimens embodied the connection between Nazi-era neuroscience and an official policy of mass murder that led to the deaths of over seventy thousand people from psychiatric hospitals across Germany and Austria from 1939 to 1941. Killing took place as the culmination of a sustained Nazi campaign to present mental patients as subhuman beings who drained the nation's resources. The victims' mental illness was taken to be intrinsic and incurable, rooted in heritable biological properties that corrupted the German gene pool. Among those who benefitted from this massacre of innocents was Julius Hallervorden, a Nazi neurologist who studied the victims' brain samples with the aim of relating mental illnesses to brain characteristics that rendered people unworthy of life. Hallervorden was reputed to have examined the brains of almost seven hundred euthanasia targets, and there is a high probability that some of the subjects were executed on demand to fuel studies.

That Hallervorden's brain samples remained in academic collections decades after World War II was a deep embarrassment to the institutions that had kept them. A memorial erected at the burial site carried a warning to scientists who transgress ethical boundaries:

> Displaced, oppressed, maltreated,
> Victims of despotism or blind justice,
> They first found their rest here.
> Science which did not respect
> Their rights and dignity during life
> Sought even to use their bodies after death.
> Be this stone a reminder to the living.

From Holmes to Hallervorden, the examples we have seen here show that diseases of the brain are not treated like diseases of any other organ—they remain as deeply personal and prone to disparagement as the disembodied conceptions of insanity that dominated in the past. The redefinition of mental illnesses in terms of brain diseases—however scientifically accurate and well-intentioned—can clear the way for cold-eyed discrimination based on neurobiological factors. Because brain problems are seen as more immutable than moral failures, brain-based discrimination can be even more pernicious than the moral condemnation that confronted previous generations of psychiatric patients. Facing cancer or cardiovascular diseases does not challenge one's sense of self-worth in the ways that facing a potentially brain-based mental illness does. And the Nazis did not exterminate people with metabolic diseases or autoimmune disorders; they killed people with schizophrenia and learning disabilities. To the extent that we reduce people to their brains and treat brains differently from other organs of the body, the stigma of the broken brain will still cut deeper than any other mark of shame that society imposes on a diseased or nonconforming individual. We shall see in the next sections that this new stigma is not only an ethically questionable response to people with mental illness but also in some cases a scientifically unjustifiable simplification of what classifies someone as mentally ill in the first place.

To the Victorian sensibilities of mid-nineteenth-century England, the natural response to the shame of mental illness was to get it out of sight. This reflex, combined with increasing diagnosis of mental pathologies themselves, led to a burgeoning of the institutionalized population from ten thousand or so in 1800 to about a hundred thousand by the end of the century. Many asylums were built to house this growing community, but the demand for facilities nevertheless greatly outpaced the supply. The mental hospitals of the day were paradoxical places where the goals of humane care came into daily conflict with the demands of an overburdened system. Stately neoclassical or gothic revival architecture, beautiful tiled hallways, and sometimes even spacious ballrooms created an ambience of gentility in striking contrast to the overcrowded conditions of the wards themselves (see Figure 12). The facilities were overseen by gentleman doctors called *alienists*, from the French word *aliéné* ("insane"), which evokes the flight of the psyche from the afflicted body. But many treatment practices were decidedly

FIGURE 12. Views of the nineteenth-century asylum: (top) ballroom of Claybury Asylum, circa 1893; (bottom) man in a restraint chair, West Riding Lunatic Asylum, 1869. *Both images courtesy Wellcome Library, London.*

alien to polite society, from the widespread use of manacles, leg irons, straightjackets, and padded cells to the drugging of patients with bromides, a discontinued family of sedatives now known to cause toxic side effects. The ideals of Tuke and Pinel, who advocated fostering moral discipline in the mentally ill, dwelt side by side with much more brutal forms of discipline. A haunting 1869 photograph from the famous West Riding Pauper Lunatic Asylum, for instance, shows an old man dressed in prisoner's stripes, strapped to a chair by his arms and neck, wincing as a head brace presses down on his crown.

To modern eyes, another anomalous feature of nineteenth-century asylums is the proportion of patients whose illnesses seemed to originate from bodily and environmental sources. Whereas most mental patients hospitalized today are people with schizophrenia, bipolar disorder, or severe depression—all conditions often cast as brain diseases—records from the nineteenth century reveal a range of peculiar extracerebral causes for commitment, of which some of the most common were financial difficulties, intemperance, and masturbation for men, or domestic trouble, "feminine problems," and childbirth for women.

The most devastating condition in Victorian mental institutions was a sickness called *general paresis of the insane*, a progressive dementia and loss of motor control that takes place in late stages of the sexually transmitted bacterial disease syphilis. It was this malady that felled the composer Robert Schumann, as we saw in Chapter 5. In an 1826 report, the French psychiatrist Louis-Florentin Calmeil chillingly described the disease as a march of symptoms in which "delirium increas[es], the reason disappears, emotional feelings are lacking, and the patient does not recognize even those around them. This gradual decline of the higher faculties, is interspersed with paroxysmal phases of variable duration, so the delirium doubles and agitation becomes extreme." By one account, up to 20 percent of British asylum inmates were admitted with general paresis, and the disease persisted at epidemic levels until the development of antibiotics in the early twentieth century. On the Continent, as well as in parts of the New World, another widespread affliction among asylum inmates was *pellagra*, a deadly syndrome associated with the "three Ds" of dermatitis, diarrhea, and dementia. One of the worst outbreaks of pellagra took place in the former US Confederate states, where it is estimated to have affected a quarter of a million people around the turn of the twentieth century. A Hungarian-American epidemiologist named Joseph Goldberger discovered eventually that the disease was caused not by a biological entity but by a dietary insufficiency of vitamin B_3.

The fact that mental illnesses of the nineteenth century could be cured by antibacterial drugs and dietary supplements is remarkable in two ways. On the one hand, the unambiguously biological nature of mental disorders caused by syphilis and vitamin B_3 deficiency—each of which promotes degeneration of neurons in the central nervous system—provides a proof of the physiological basis of the human mind. On the other hand, the etiologies of general paresis and pellagra argue against equating mental illnesses with diseases of the brain per se. Paresis and pellagra act through the brain but not because of it. Pathologies like these instead exemplify the multilayered context in which instances of mental illness can emerge. There is the cognitive and behavioral deficit itself, as well as the abnormal brain biology that accompanies it. There is also a wider network of causative factors, including not only the specific stimuli or pathogens that induce brain dysfunction but also the social and environmental milieu in which such factors propagate. Thus, a condition like general paresis can simultaneously be a brain disease, a bacterial illness, and a social pathology. This is a brand of complexity highlighted by the Polish physician and medical historian Ludwik Fleck, who described the amalgamation of diverse moral and medical views that came to define syphilis itself in the premodern era.

If the mental illnesses of the past were caused by a balance of external and internal influences, then perhaps some of today's mental illnesses are also similarly "delocalized." We considered in Chapters 6 and 7 the extent to which factors outside the brain affect neural function and behavior. Surely it would not be surprising if such a mix of factors also contribute to pathologies of neural function and behavior. In fact, there is well-known evidence for both internal and external influences on mental health, resulting in a causal web that defies any simple equivalence between mental illnesses and brain diseases.

Although we do not know enough about the biological basis of most mental disorders to determine "how much" of each condition might be innately determined versus environmentally caused, we can get an approximate idea by looking at the extent to which each mental illness appears to be inherited by children from their parents. If an illness is inherited in the same way as traits like dark hair or short stature, it means that the condition is caused mainly by factors that are written into our genes and present in our DNA from the moment of conception. Correlating mental illnesses with genetic data therefore provides a way to determine the extent to which an illness is heritable.

If a person with a mental illness has an identical twin, the simplest way to check for a genetic link would be to see if the twin also has the disease. Because identical twins share 100 percent of the same genes, they should also share traits like depression or schizophrenia if these disorders are genetically determined. For the majority of people with mental illness who do not have identical twins, we could still check for genetic causes by seeing if they have close family members who also have the illness. Because each person has a high percentage of genes in common with their parents, siblings, and children, one would expect a greater than average likelihood that close relatives share disease characteristics, provided the disease is genetically determined. With advances in genetic mapping technology, another powerful way to test whether a disease has a heritable component is to collect genetic data from thousands of people both with and without the illness, and then examine the extent to which the presence of the illness correlates with the presence of any particular gene variant or set of variants. In any of these types of studies it is of course important to make sure that extraneous factors do not bias the results. For instance, in studies of genetic factors in twins, it is important to distinguish shared environmental influences from the influences of the twins' shared genetic inheritance. Researchers do this by establishing comparison groups, such as nonidentical twin siblings for twin studies, or adopted relatives for family studies, in which environmental factors would be the same but genetic makeup would be different.

Using approaches like these, scientists have calculated the fraction of disease prevalence that can be explained by variations in genetic makeup—a quantity called *heritability*—for many mental illnesses. Instances of a disease with a heritability of 1 are explained entirely by genetic variations, whereas occurrences of a disease with a heritability of 0 are thought to be explained entirely by environmental factors. Based on reputable data from multiple studies available as of 2012, geneticists Patrick Sullivan, Mark Daly, and Michael O'Donovan reported heritabilities of several of the major mental illnesses. Leading the list was schizophrenia, with a heritability of 0.81; at the bottom was major depressive disorder, with a heritability of 0.37. Conditions including bipolar disorder, attention-deficit hyperactivity disorder, nicotine dependence, and anorexia nervosa showed intermediate values. The upshot from these findings is that both genes and environment seem to contribute to major mental illnesses, suggesting that inborn biological factors are not on their own enough to explain these diseases.

There are subtleties to how the heritability statistics should be interpreted. For instance, the fractional heritability of a disease like major depression could mean that instances of the disease generally correlate weakly with a variety of genetic factors, or it could mean that a fraction of occurrences is perfectly linked to specific genes while another fraction is not explained by genes at all. Notably, heritability does not reflect specifics of how the diseases manifest themselves, and in no case does it predict what will trigger individual episodes of illnesses like depression, bipolar disorder, and schizophrenia. Also importantly, the existence or absence of genetic links for a mental illness neither rules in nor rules out a causative role for brain biology in the disorder. Genes that correlate with a disorder could affect the brain only indirectly, for example, by altering components of emotion-related physiology that lie outside the brain, or by shaping a person's appearance in a manner that influences his or her social status. Conversely, the absence of a clear link between genes and mental illness does not mean that a brain pathology is not central to the condition. For instance, traumatic brain injuries often induce mental complications without involving genetic causes in direct ways. In any illness that affects behavior the brain must be involved, but whether brain abnormality is an underlying cause or a second-order effect cannot be determined based on genetics alone. Although the genetic evidence suggests that the environment is a major contributor to mental illnesses, to understand the nature of possible environmental roles, we need to look at the wider world.

In the 1930s, two sociologists at the University of Chicago set out to do just this. The university at that time was host to the emerging Chicago School of sociology, a movement of activist scholars who fanned out into the urban neighborhoods around them and documented the patterns of living they saw there. One such scholar was Ruth Shonle Cavan, who in 1928 had published a groundbreaking study that found increased suicide rates in "communities where social disorganization prevails." A young graduate student named Robert Faris was inspired by Cavan's work and wanted to apply her approach to a survey of mental illnesses. He teamed up with H. Warren Dunham, a fellow student, and together the two of them examined thirty-five thousand cases of mental disorders in the Chicago

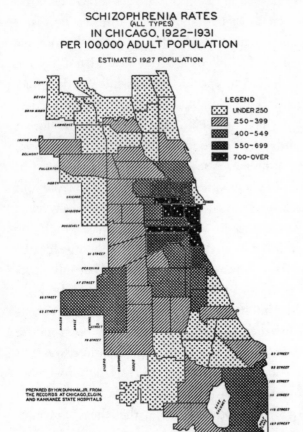

FIGURE 13. Map of schizophrenia statistics in Chicago, as diagrammed by Robert Faris and H. Warren Dunham in 1939.

area over a thirteen-year period. These data revealed a striking phenomenon: the incidence of schizophrenia, the most mysterious of the mental disorders, correlated closely with the urban environment (see Figure 13).

The number of schizophrenia cases peaked in the downtown slums, near where Anish Kapoor's reflective Cloud Gate now sits, and declined smoothly as distance from this epicenter increased. In the leafy residential neighborhoods like Highland Park in the north or Hyde Park in the south, schizophrenia rates were only about 20 percent as high as those in the center. The result was apparently unrelated to the race or nationality of people who lived in different areas of Chicago. It was also specific to schizophrenia—mental diseases such as depression and bipolar disorder did not show this pattern. Faris and Dunham's results were published in their 1939 book, *Mental Disorders in Urban Areas*, which is considered a landmark of psychiatric epidemiology to this day.

Toward the middle of the century it became common to relate mental pathologies to social pressures and problems. Faris and Dunham were criticized for giving in to this bias and for ignoring genetic data about the origins of schizophrenia that already existed at the time. But subsequent studies have clarified and confirmed aspects of their findings. For a start, similar correlations between urban dwelling and schizophrenia turned up subsequently in numerous cities in multiple countries. Although some hypothesized that the effect could be explained by "social drift" of psychotic people into urban zones where standards of living were lower and drug abuse more prevalent, experiments in Europe showed that the simple fact of being born or raised in an urban environment carried extra risk of developing schizophrenia later in life. There is still no single convincing explanation for the consistent correlation of urban living with schizophrenia, but the very fact that this relationship exists is eyeopening evidence of the influence of environmental input on this particularly devastating condition.

Epidemiologists have found many other intriguing relationships between environmental variables and mental illnesses. Schizophrenia itself is correlated with ethnic minority status, especially for African or Afro-Caribbean immigrants and their descendants in white-majority countries, but not in their countries of origin. Marijuana and other illegal drugs seem to increase the risk of developing schizophrenia approximately twofold. Finally, being born in winter—regardless of which hemisphere— notably increases the probability of becoming schizophrenic, suggesting a connection to seasonal infectious diseases. Major depression, meanwhile, is higher among unemployed people, including homemakers who are not seeking employment. Divorced, separated, and widowed individuals have double the rates of depression of married or nevermarried people, even when controlling for differences in age and sex. Bipolar disorder, which shares characteristics with both schizophrenia and depression, displays a blend of associations, including elevated rates among people with low income and education and also among divorced and widowed individuals.

Perhaps we do not need epidemiologists to tell us that mental disorders are at least partially products of the world around us. The writer Elie Wiesel mused that "within each of us, whether in good health or bad, there is a hidden zone, a secret region that opens out onto madness. . . . One misstep, one unfortunate blow of fate, is enough to make us slip or flounder with no hope of ever rising up again." We all know stories from literature that reinforce Wiesel's narrative. In Sylvia Plath's semi-autobiographical novel, *The Bell Jar*, the heroine's descent into depression is triggered by rejection from a writing program she had applied for; this compounds her preexisting feelings of professional and personal disappointment. Dostoyevsky's Raskolnikov meanwhile flits in and out of sanity as he grapples with the consequences of the terrible murder he commits in *Crime and Punishment*. Most famously, Shakespeare's King Lear is driven mad by the callousness of his daughters—it is not his own genes but those who inherited them that send him wandering mindlessly on the heath.

Among environment, genetics, and the biology that comes between, the causes of mental illnesses may be just as complex as the mystified brain itself. The diseased brain has been compared to a broken car, as we saw earlier, but a mental illness is more like a car accident in which several elements conspire. The accident could be the combined product of problems with the car, the driver, and the road itself, just as a psychiatric disorder represents the influences of instantaneous brain function, genetically determined predispositions, and the environment and society more broadly. In some cases, a car might simply not have been built for the driving conditions it was used in, just as a mentally ill person might not have been equipped to handle a particular set of events or conditions that triggered a breakdown. Evidence for multifactorial causation of psychiatric conditions comes as no surprise to people who work with mentally ill patients or who have experienced mental illness themselves. But again, the cerebral mystique tends to distort the picture. If we focus too much attention on the brain itself, we lose sensitivity to the context that makes the brain do what it does. This again ignores the cardinal lesson of neuroscience: that the brain is a biological organ of flesh and blood, embedded in a continuum of causes and effects. There is an even deeper reason not to reduce mental illnesses to pathologies of the brain alone, however, and that has to do with the concept of mental disease itself.

On July 7, 1970, a Russian poet named Natalya Gorbanevskaya stood trial in the basement of a Moscow city court. She had been arrested a few months earlier, shortly after self-publishing an un-sanctioned account of a demonstration she and seven other dissidents had held against the 1968 Soviet invasion of Czechoslovakia. Now she was accused under Article 190-1 of the Russian Criminal Code, which defined the crime of "dissemination of fabrications known to be false, which defame the Soviet political and social system." No more specific charges were brought, but Gorbanevskaya could not have defended herself against them in any case. She was at that time locked away in the notorious Butyrka prison, which had once housed the KGB founder, Felix Dzerzhinsky himself. Furthermore, doctors had examined Gorbanevskaya and determined that she was medically unfit to take part in the proceedings. Experts at the Serbsky Institute, Russia's premier center for forensic psychiatry, certified the defendant "to be of unsound mind and to require forcible treatment in a psychiatric hospital of special type."

Daniil Lunts, head of the Serbsky's diagnostic section, provided the most important evidence at the trial. He testified that Gorbanevskaya "was suffering from a creeping form of schizophrenia which 'has no clear symptoms' but which causes a change in the sphere of the emotions and the will, in thought patterns, and in an insufficiently critical attitude toward one's own mental state." With no overt evidence to address, defense lawyer Sofia Kalistratova could hardly contest Lunts's professional judgment. As a prominent advocate of dissidents, Kalistratova recognized the grim nature of his findings, however. *Sluggish schizophrenia* was the invention of Andrei Snezhnevsky, director of the Institute of Psychiatry at the USSR Academy of Medical Sciences and the dominant figure in Soviet psychiatry at the time. Diagnosis with this phantom disease was the basis for involuntary institutionalization of many who protested the Soviet regime. Victims were locked up indefinitely in mental hospitals around the country, where they could be isolated, beaten, and forcibly medicated. Faced with the prosecution's predetermined case, all Gorbanevskaya's team could do was plead. "If my daughter has committed a crime, sentence her to any punishment, even the most severe," begged the defendant's mother at the hearing, "but do not place an absolutely healthy person in a psychiatric hospital."

Gorbanevskaya served two years in medical detention, where she was drugged and went on a brief hunger strike. She was released in 1972 and soon after fled to France, where she dwelt until her death in 2013. The popular folksinger Joan Baez performed a song in her honor, containing the lyrics, "Are you insane / As they say you are / Or just forsaken . . . I know this song / You'll never hear / Natalia Gorbanevskaja." Although Gorbanevskaya had been lucky enough to escape the song's prophecy and spend her later years in relative comfort, hundreds of other Eastern Bloc dissidents were less fortunate. The practice of politically motivated psychiatric incarceration continued until the fall of the Soviet Union in 1991, and is said to continue in other parts of the world even now.

As we look back with horror on the abuses of Soviet psychiatry, we judge doctors like Lunts and Snezhnevsky in the harshest terms. A Russian émigré at one point described Lunts as "no better than the criminal doctors who performed inhuman experiments on the prisoners in Nazi concentration camps." But Walter Reich, a professor of psychiatry and behavioral sciences at George Washington University, has suggested differently. In 1982, he had travelled to the Soviet Union and personally interviewed Andrei Snezhnevsky. In a *New York Times* piece written shortly after the visit, Reich offered the view that "because of the nature of political life in the Soviet Union and the social perceptions fashioned by that life, dissenting behavior really does seem strange there; and that, because of the nature of Snezhnevsky's diagnostic system, this strangeness has, in some cases, come to be called schizophrenia." In other words, perhaps the Serbsky Institute doctors were acting in good faith when they diagnosed Gorbanevskaya and other political prisoners with sluggish schizophrenia—perhaps, as Reich puts it, they "really believed that the dissidents were ill."

The very fact that Reich's hypothesis can be entertained speaks to a fundamental flaw in the notion that all mental illnesses are brain diseases: the concept of mental illness itself is essentially subjective. Unlike many diseases, where there is a germ, a suppuration, a growth or lesion that can be found or not found, the diagnosis of most mental illnesses hinges on the opinion of the professional making the judgment and on the communally determined standards that person has to work with. Where once such standards were informally set by cultural expectations of reasonable conduct or moral behavior, today in the United States the specifications of mental health are codified into a bible of the psychiatric profession: the American Psychiatric Association's *Diagnostic and Statistical Manual* (DSM). In its fifth and latest edition, the DSM was put together by an international team of over 160 mental and medical professionals, in consultation with more than 300 advisors representing myriad professional groups and communities, as well as commentators from the public who provided extensive feedback. DSM-5 provides a list of criteria for about three hundred different psychiatric conditions, each decided by consensus building among the various individuals and groups who gave input.

Although there has been broad agreement about the essential features of some psychiatric conditions since the first DSM came out in 1952, the boundaries of most illnesses remain vague. For instance, the DSM guidelines for diagnosis of schizophrenia require a patient to display two of the following symptoms within a one-month period: delusions, hallucinations, disorganized speech, disorganized behavior, and negative emotions. The criteria continue, "For a significant portion of the time since the onset of the disturbance, level of functioning [must be] below the level achieved prior to onset." Clearly the time window of one month is arbitrary, and it is up to the diagnosing doctor to identify the moment of disease onset and judge the severity of symptoms. (When does fantasizing about becoming a movie star cross the line from ambition to delusion, and when does speaking with the dead become a hallucination rather than a spiritual experience?) The shifting of disease categories over the past sixty years provides further testament to the social factors involved in defining mental illnesses. The 1952 DSM listed only 106 disorders, only a third the number recognized in DSM-5. Along the way, entries for neurosis and homosexuality were dropped, while new categories for autism and attention deficit disorders were introduced. Although some of these changes could be attributed to new scientific knowledge, others simply reflect cultural shifts. At the end of the day, the definition of mental illness is largely statistical: the mores of the majority in any given place and time determine the limits of sanity.

By looking around the globe, journalist Ethan Watters has seen firsthand how variable perceptions of mental pathology can be. He describes exotic mental conditions like *amok*, a Malay term for a sudden outburst of violence or suicidal behavior, and *zar*, a mood disorder of Middle Eastern women that is traditionally exorcised during a ceremony of ecstatic dancing and singing. Watters notes that cross-cultural surveys reveal "an impressive body of evidence suggesting that mental illnesses have never been the same the world over (either in prevalence or in form) but are inevitably sparked and shaped by the ethos of particular times and places." He also observes an encroachment of American psychiatric practices on other cultures, seen most prominently in the adoption of American disease categories in other countries. With that encroachment, DSM-defined mental illnesses appear to be replacing earlier aboriginal maladies. As Watters describes it, "a handful of mental-health disorders—depression, post-traumatic stress disorder and anorexia among them—now appear to be spreading across cultures with the speed of contagious diseases."

The culturally influenced nature of psychiatric disease categories means that any genetic or neurobiological features that correspond to the categories also reflect cultural biases. If Soviet geneticists had found that certain genes correlate with sluggish schizophrenia in political dissidents, then possession of these genes would be considered a risk factor for sluggish schizophrenia in others who might not yet have "come out" as dissidents. Biologists could study what it is about the relevant genes that makes people sluggishly schizophrenic. Using the latest molecular technologies, they could even make mouse models of the disease by altering the genes of mice to mimic those in human dissidents; then these mouse models could be exhaustively studied by a battery of biological techniques. Such hypothetical studies would parallel work that many real-life scientists now perform with DSM-listed conditions, and they might even come up with answers. It is quite possible, for instance, that genetically associated personality characteristics dispose people to becoming dissidents and displaying whatever psychological characteristics doctors like Snezhnevsky and Lunts said they were detecting—but would these genes and any related neurobiological phenomena truly underlie a "brain disease"? The genetic and physiological correlations with dissident behavior might be real enough, but the characterization of these as disease-related *defects* would remain just as subjective as when the Serbsky doctors first made their diagnosis.

In 1960, a psychiatrist named Thomas Szasz laid out a radical response to this conundrum in an essay he provocatively entitled "The myth of mental illness." Szasz explained that "the notion of mental symptom is . . . inextricably tied to the *social* (including *ethical*) *context* in which it is made." Meanwhile, he advised that "for those who regard mental symptoms as signs of brain disease, the concept of mental illness is unnecessary and misleading." In other words, if there is genuinely something identifiably wrong with a person's brain, then there is no added value to discussing it in mental terms. Instead, Szasz regarded mental illnesses that do not correspond to obvious brain abnormalities simply as "problems in living," which should not be medicalized and treated with drugs or hospitalization. Szasz earned notoriety as a heretic in his day, and many of his fellow psychiatrists regarded his denial of mental illness as an irresponsible attack on the profession, but others saw him as a praiseworthy advocate for patients against excesses of the medical establishment. Whether or not one sympathizes with him, however, Szasz's critique perfectly highlights one of the most important consequences of how we regard mental diseases: our understanding of mental illnesses guides the ways in which we choose to treat them.

We have seen that mental illness is a multilayered phenomenon; environmental and cultural factors interact with intrinsic human biology to create both the manifestations and perceptions of disease. To the extent that we focus exclusively on one level versus another, our attitudes about appropriate therapies are distorted. For instance, if back in the nineteenth century we had come to regard general paresis first and foremost as a brain disorder, it would probably not have occurred to us that an infectious bacterium could cause the disease, and we would not think to use antibiotics. Instead we might turn to drugs that directly counteract syphilis-induced brain degeneration. These could include so-called neuroprotective agents—substances including caffeine, fish oil, and vitamin E are examples—that help reduce neurodegeneration. Alternatively, if we chose to regard the disease as a result of dissolute morals, as indeed was common before the age of modern medicine, our tendency might be to teach patients better practices, encourage a monogamous lifestyle, and regulate prostitution.

The question of how best to analyze mental illnesses is in fact one of the most hotly contested topics in medicine. Two poles of the debate were defined in an influential 1977 article in *Science* magazine by George Libman Engel, a psychiatrist at the University of Rochester. Engel described what he termed the *biopsychosocial model*, wherein appreciation for biological aspects of disease is blended with consideration of psychological and social factors in how the patient experiences symptoms and responds to treatment. He contrasted that with the *biomedical model*, in which disease is defined by a specific biological cause, usually at a molecular level, and treated exclusively using drugs, surgery, and other medical technologies. Engel was not neutral in the debate; as a young physician in the 1940s he had become immersed in a field known as psychosomatic medicine, which focuses on the interaction between social and psychological factors with bodily functions. Defending his specialty against the advance of molecular medicine and the increasing dominance of the biomedical perspective, Engel lamented a mindset in which doctors feel they "need not be concerned with psychosocial issues which lie outside medicine's responsibility and authority." Doing this, he wrote, would either "exclude psychiatry from the field of medicine" or limit its reach to "behavioral disorders consequent to brain dysfunction."

The contrast between Engel's two poles maps to some extent onto the two dominant forms of psychiatric treatment today: talk therapy versus drugs. Talk therapy has many different flavors, but each places emphasis on helping patients cope with problems at a psychological or social level. Examples include psychoanalytic approaches, which focus on discerning unconscious or forgotten thoughts and their relevance to emotional disturbances, as well as more modern techniques like cognitive behavioral therapy, which focuses on training patients out of counterproductive habits of mind and action. Pharmacological treatments, by contrast, directly alter physiological processes in the brain. Many drugs target specific neurochemical processes; examples include serotonin reuptake inhibitors for depression and benzodiazepines that act as sedatives by mimicking the brain's main inhibitory neurotransmitter, GABA. Other psychotherapeutic drugs act by uncertain mechanisms, like lithium for bipolar disorder, but are nevertheless widely prescribed.

Recent surveys of health care practices reveal clear trends toward increased use of medicines and decreased use of psychotherapy. These statistics illustrate the continuing shifts in medical culture that Engel was already protesting forty years ago. A widely cited report by Medco Health Solutions documented consistent increases from 2001 to 2010 in the percentages of US men, women, and children using mental health medications. During this time, the percentage of adults using antidepressants or the latest generation of antipsychotics roughly doubled. A similar study in England found that prescriptions of drugs for mental disorders increased by about 7 percent per year from 1998 to 2010. Meanwhile, a study in the *American Journal of Psychiatry* reported that the proportion of patients receiving psychotherapy for mental health conditions declined from 56 percent to 43 percent from 1998 to 2007, while the proportion of patients receiving drug therapy rose from 44 percent to 57 percent.

It is difficult to prove that increasing use of psychiatric medications results from increasing belief in the causative role of brain biology, but such a relationship seems likely. We know that awareness of neuroscience has been on the rise over the same period during which the popularity of psychopharmacology increased. The same surveys that reported the correlation between trusting brain-based explanations and stigmatizing mental illness also found that neuroscience literacy correlates with comparatively more openness to psychiatric medicines. To the extent that people believe that conditions like schizophrenia or depression result from neurochemical imbalances or other brain dysfunctions, it certainly makes sense that they would be more apt to treat their condition using drugs that act directly on the brain. Psychologists Sally Satel and Scott Lilienfeld accept this logic but take issue with the result. They write that "the brain-disease model leads us down a narrow clinical path," but that "it overemphasizes the value of pharmaceutical intervention." Medical journalist Robert Whitaker goes further, blaming the increased use of medication for an epidemic of psychiatric side effects. Whitaker thinks that the public is being fed a false narrative about the efficacy of psychopharmacology. He says that patients are taught from an early age that "they have something wrong with their brains and that they may have to take psychiatric medications the rest of their lives, just like a 'diabetic takes insulin.'"

But the dichotomy between medicine for the brain and talk for the mind is a false one, reinforced in part by the cerebral mystique. Even if we accept the brain's central role in our mind and behavior, it should be easy to understand that a variety of internal and external ways of interacting with the brain could help a troubled individual. Conversely, we cannot ignore the fact that psychotherapies act not on a metaphysical soul but on a physical person, who might benefit just as readily from an empathetic voice as from an effective medicine. "The biological/psychosocial treatment divide between pharmacotherapy and psychotherapy is a myth because the target of both therapies is diseased neural functioning," explain Aaron Prosser, Bartosz Helfer, and Stefan Leucht in a recent editorial in the *British Journal of Psychiatry*. They point out that the difference lies simply in the mode of action: drugs produce relatively nonspecific changes to brain chemistry, while talk therapy provides "tailored modulation" of the same biological phenomena. This conclusion is important because it may counteract the notion that nonpharmacological treatments for mental disorders are inherently nonscientific, and thus helps make it easier for patients to get the therapies they will gain the most from.

The dichotomy between pharmacotherapy and psychotherapy might also be false for a different reason: neither of these avenues approaches the problem of mental illness at the level of communities and cultures. Instead, both drug and talk therapies share a narrow focus on the minds or brains of single patients, reflecting a similarity between old and new. But the connections among brain, body, and environment imply that detecting and treating mental disorders are important at a level that goes beyond the internal lives of individuals. History tells us that the classic disorders of general paresis and pellagra would never have been eliminated had it not been for epidemiological studies and the public health efforts that followed them. We also know from phenomena like the persistent correlation of schizophrenia with urban birth or the association of bipolar disorder with low income and education that there is much more to be discovered about how context contributes to psychiatric issues. For this reason, we cannot think of mental illnesses merely as problems with individual brains or minds; we must view each sick person as part of the context they inhabit, in which social and environmental forces act in parallel with biological factors to influence mental health. Lessening the burden of mental illness may depend on effectively addressing these contextual forces.

Viewing mental illness as a problem not just of individuals but of their circumstances may also be one of the most compelling ways to reduce stigma. Patrick Corrigan and Amy Watson refer to such contextualized accounts of psychiatric disorders as *psychosocial* explanations, omitting the *bio-* from George Engel's earlier term. "Instead of arguing that mental illness is like any other medical illness," they explain, "psychosocial explanations of mental illness focus on environment stressors and trauma as causal factors." Citing several studies, Corrigan and Watson write that "in contrast to biological arguments, psychosocial explanations of mental illness have been found to effectively improve images of people with mental illness and reduce fear." Rather than dehumanizing patients by blaming their broken brains, and rather than delegitimizing patients by condemning their moral deficiencies, psychosocial explanations instead depict mental illnesses as "understandable reactions to life events."

Modern technology may bring increased capability for discerning and treating social and environmental aspects of mental illness. An indication of this potential came in 2015, when Thomas Insel, then director of the US National Institute for Mental Health, stunned his colleagues by announcing that he was leading the government to join the tech giant Google. In explaining the move, Insel described his dream of improving diagnosis and therapy using the power of internet-based approaches. "Technology can cover much of the diagnostic process because you can use sensors and collect information about behavior in an objective way," explains Insel, hinting that such measurements might come to supplant DSM-guided diagnoses. With networked sensors, one could distinguish subtle differences in a person's voice or behavior that might give indication of a brewing mental problem. "Also, a lot of the treatments for mental health are psychosocial interventions, and those can be done through a smartphone," Insel adds. He suggests in particular that cognitive behavioral therapy, a form of psychotherapy inspired by behaviorist training methods, could be implemented remotely via people's electronic devices, reducing the barrier for obtaining treatment.

Much of the excitement around internet-based approaches in psychiatry revolves around the idea of detecting mental illnesses by monitoring online behavior. For instance, a depressed person could be identified and diagnosed based on patterns of usage, or perhaps the content of their online activities. Psychologists Adrian Ward and Piercarlo Valdesolo write that "peer-to-peer file sharing, heavy emailing and chatting online, and a tendency to quickly switch between multiple websites and other online resources all predict a greater propensity to experience symptoms of depression." It could be possible to use data of this kind to identify people who need medical attention, although this also evokes blatant privacy concerns. At a higher level, however, lies the less controversial idea of relating internet-detectable signs of pathology anonymously to social and environmental contexts such as neighborhoods, economic factors, cultural niches, and so on. In this framework, the internet-based screening would serve as a sort of mental health census, with a level of sensitivity to multiple criteria that no existing survey has provided to date. This could constitute a modern form of the classic studies in psychiatric epidemiology, and it could reveal with remarkable clarity how mental health is a phenomenon not just of individual broken brains but of the much broader mental habitats humankind occupies across the globe.

Technology can be a vehicle for understanding the brain's input and output relationships in both health and disease. By illuminating the causal fabric that surrounds each of us, technology in this role might help counteract the view that our brains are merely physical substitutes for the autonomous souls we once believed we had. Technology could also become a vehicle for purposefully altering brains and brain function—possibilities that for decades already have fed a diverse array of futuristic visions. In the next chapter we will see how the cerebral mystique motivates but also constrains these visions.

nine

NEUROTECHNOLOGY UNBOUND

T HE FIRST SUPERMAN GOT HIS POWERS FROM BRAIN enhancement. Before there was Kal-El, the strange visitor from another planet who arrived on Earth with abilities far beyond those of mortal men, there was Bill Dunn, a destitute earthling who became transhuman by consuming a man-made drug. During the depths of the Great Depression, Dunn is plucked from a breadline by an unscrupulous chemist named Smalley. Lured by the promise of a square meal, the vagabond goes off to the scientist's house but finds his drink laced with a psychoactive potion Smalley has recently developed. Dunn becomes dizzy and delirious, but soon recovers and realizes that he has gained supernatural powers of telepathy and clairvoyance. "I am a virtual sponge that absorbs every secret ever created," the newly minted Superman declares. "Every science is known to me and the most abstruse questions are mere child's play to my staggering intellect. I am a veritable God!"

The Dunn-Superman learns how to take advantage of his new talents, but many of his schemes come at the expense of others. He injects into people's minds the desire to donate their money to him. A drugstore clerk gives him ten dollars without questioning, and a rich tycoon later cuts Dunn a check for $40,000 (equivalent to $700,000 today) without even having met him. Given his ability to see into the future, the Superman also turns out to be a very successful investor. As he becomes more confident in his powers, he also grows more destructive, however. He slays Smalley and tries to set off global conflict by sparking a diplomatic showdown at a fictionalized League of Nations. He is on the brink of murdering an unfortunate journalist sent to investigate him, when without notice the potion suddenly starts to wear off. Dunn metamorphoses back into the wretched hobo he started as, and his last prophetic vision is that of himself back in the breadline.

The strange story of Bill Dunn appeared in 1933 as a nine-page illustrated magazine piece entitled "The Reign of the Superman." It was composed and self-published by two high school students named Jerome Siegel and Joe Shuster. Two years later, the pair reworked their initial Superman into the Man of Steel we now know and love; they sold the concept to Detective Comics in 1938, and the rest is history. As the superhero from Krypton went on to fame and glory, the more mundane prototype was forgotten. But the two teenagers' early tale of an ordinary man made extraordinary by brain technology is one that now profoundly resonates with the hopes and fears of our later age—an age in which futuristic technology for modifying and manipulating the human nervous system is increasingly finding its way into reality.

Among the wonders of today's neurotechnology are pills for making people smarter, devices that remotely monitor or stimulate the nervous system, and genetic techniques that could reshape the structure of the brain itself. One can easily imagine some of these tools in the arsenal of a comic book hero or villain, and Siegel and Shuster's story gives us a taste of how things can begin to go awry; the hazards of human experimentation and the unscrupulous exploitation of technology for personal gain and injury to others are just some of the dangers. In the real world, we must think carefully through how any new neurotechnology can be safely and ethically applied. We must also decide what future technologies we should work toward and to what end. Should we strive to create genuine supermen or guard ourselves against their emergence?

In this chapter we will consider how the cerebral mystique and its idealization of the brain affect thinking about neurotechnology. We will see how the mystique adds to the allure of artificial brain interventions but also fosters artificial distinctions between technologies that act directly and indirectly on the brain. A more down-to-earth view of the brain and its relationship to the body and environment might degrade these distinctions and change the way we approach neurotechnology and its development; just as importantly, it could prompt us to look more closely at social issues around some of the less ostensibly neural technologies that manipulate our minds.

Perhaps nothing better exemplifies the promises and perils of neurotechnology than the concept of *hacking the brain*, a meme that in recent years has proliferated across the popular media. Aspirations and anxieties around this idea reflect the prevalent but questionable notion that purposefully changing people's brains could be a pertinent way to change their lives. In a 2015 *Atlantic* article that takes "Hacking the Brain" as its title, journalist Maria Konnikova ties the phrase explicitly to the futuristic goal of enhancing intelligence. Other writers emphasize efforts in the here and now to alter human behavior by manipulating the brain with electrical or magnetic stimulation. Numerous talks in the trendy TED lecture series reference some form of brain or mind hacking, from neurosurgeon Andres Lozano's presentation on how hacking the brain can make you healthier to the talk by magician Keith Barry, who we should "think of . . . as a hacker of the human brain." The tone of these reports is typically exuberant—"Need more proof that the future is here?" asks the TED website. But some adopt a more apprehensive angle. Konnikova, for instance, wonders whether brain hacking could also lead to a "dystopia where an individual's fate is determined wholly by his or her access to cognition-enhancing technology" or where "some Big Brother–like figure could gain control of our minds."

Hacking is a living word with connotations that mirror this ambiguity. For me, the primary associations are with machetes, cleavers, scythes, and their use in such contexts as butcher shops, jungle forays, and the Rwandan genocide. For the students I teach at MIT, the dominant definition is a digital one, however. At a university where five of the ten most popular classes are in computer programming, hacking refers most often to the subversive but generally harmless pastime in which engineering aficionados breach and alter computer security systems, software, or electronic hardware. MIT is also famous for "hacks"—technically sophisticated pranks played by students to amaze others on and sometimes beyond campus. In this vein, MIT hackers once transported a police car to the top of the school's Great Dome; they also stole an iconic cannon from rival university Caltech and reinstalled it at MIT. These different senses of hacking appear at first to be only loosely related, but they share an association with invasion and indelicacy. For example, although hacking the iPhone operating system does not literally split anything open, it does involve forcing one's way into a previously forbidden space in the software of the device. Although the prankster's hacking is not as brutal as the slashing of a slaughtered carcass, it is often done without subtlety, using whatever means are available.

Most people probably think of hacking the brain as something closest to the digital definition—it connotes breaking into the brain and manipulating it, typically by connecting it to artificial gadgetry like electrodes or fancy scanners. The rationale for brain hacking therefore benefits from the ubiquitous brain-as-computer analogy we saw in Chapter 2. Hacking the brain can seem glamorous, perhaps because it combines technological sophistication and edginess in the manner of MIT-style pranks, but the reality is often somewhat gruesome as well. This is because brain hacking almost invariably involves some kind of attack, either through the physical violence of surgery and disruption to biological tissue or through less damaging methods like fMRI that nevertheless intrude into the brain's private places. Hacking the brain is not necessarily a good thing to do.

The most common contexts for brain manipulations are medical. For over a hundred years, physicians have used what is known as *resective neurosurgery*—the slicing apart of cerebral structures—to treat a variety of neurological and neuropsychiatric diseases, as well as brain cancers. The most infamous resective technique was the prefrontal lobotomy, a now-extinct treatment for schizophrenia introduced in the 1930s by the Portuguese neurosurgeon António Egas Moniz. The lobotomy involved

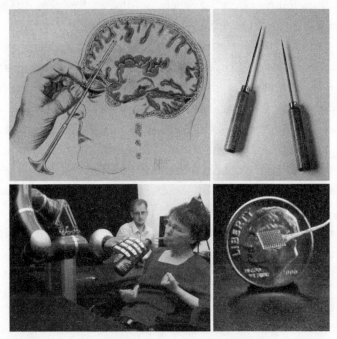

Figure 14. Hacking the brain using old and new technology: (top left) diagram of the transorbital lobotomy procedure developed by surgeon Walter Freeman, (W. Freeman, "Transorbital leucotomy: The deep frontal cut," *Proceedings of the Royal Society of Medicine* 41, 1 Suppl [1949]: 8–12, copyright ©1949 by The Royal Society of Medicine; *Reprinted by permission of SAGE Publications, Ltd.*); (top right) Freeman-style lobotomy instruments (Wellcome Library, London); (bottom left) Cathy Hutchinson controlling a prosthetic arm using her brain-machine interface; (bottom right) closeup view of the BrainGate electrode array implanted in Hutchinson's brain. *Bottom images courtesy braingate.org.*

cutting the white matter of the brain's frontal lobes, which severed neural connections between these regions and the rest of the cerebral cortex. This process could in some cases reduce symptoms of psychosis, but only at considerable risk to the person undergoing the operation. In one variant, the surgeon would hammer a long metal needle through the back of the patient's eye socket and then swipe the device sideways to carve across deep brain structures—a true brain hack if ever there were one (see Figure 14). Approximately 5 percent of lobotomy patients died during the surgery, more than one in ten developed postoperative convulsions, and many of the other survivors became impassive or catatonic. Lobotomies were nevertheless performed on thousands of patients up until the late 1960s, including celebrities such as John F. Kennedy's sister Rosemary and Argentina's first lady Eva "Evita" Perón.

Although the lobotomy fell out of favor half a century ago, closely related forms of surgical brain hacking remain in wide use today. Most prominently, resective procedures are performed each year on hundreds of epileptic patients whose seizures cannot be controlled with drugs. In cases where seizure onset is linked to a focus of pathological activity in specific brain areas, doctors can try to limit the frequency or severity of attacks by destroying the focus or cutting around it. Although fewer than 10 percent of subjects experience significant complications from modern epilepsy surgery, there have been spectacular setbacks in the past. Most famous was the case of Henry Molaison, whose left and right hippocampal brain regions were removed during epilepsy surgery in 1953, leaving him without short-term memory for the rest of his life. Molaison's experience led scientists to new insights about the role of the hippocampus in memory formation but also underscored the dangers inherent in such invasive techniques.

Modern forms of medically sanctioned hacking complement the neurosurgeon's knife with more nuanced approaches. A technique called *deep brain stimulation* (DBS) has become one of the most widely used; it is used to treat movement-related problems like Parkinson's disease and obsessive compulsive disorder, and it has now been applied in over one hundred thousand patients. DBS involves insertion of electrodes into the brain through small holes drilled into the skull. Each electrode is connected via subcutaneous leads to an implanted control module about the size of a cookie; at regular intervals, the module sends tiny pulses of electrical current through the wiring, delivering little kicks of energy to neurons near the electrode tip. Like resective surgery, DBS treatment is thought to act primarily by inactivating tissue in the neighborhood of the intervention, but it is reversible and can be adjusted as necessary. More experimental brain-hacking techniques use electrodes both to stimulate and to record signals from patients' brains. The resulting information can be used to control DBS-style treatments in real time. Brain recordings can also help paralyzed patients interact with prostheses or other external devices, via what are known as *brain-machine interfaces* (BMIs). In an amazing demonstration of this technology, neuroscientists John

Donoghue, Leigh Hochberg, and their colleagues implanted an array of ninety-six microelectrodes into the cerebral cortex of a paralyzed woman named Cathy Hutchinson. Using the BMI, Hutchinson gained the ability to control a robotic arm with her thoughts; she was able to serve herself a drink for the first time since suffering a devastating stroke fifteen years earlier (see Figure 14).

Breakthroughs such as Hutchinson's BMI ignite the imagination and drive much of the fascination with brain hacking. Controlling a mechanical device using neural activity alone sounds almost superheroic, like Wonder Woman's ability to fly an invisible airplane using only her mind. Could it be that these and other amazing powers are just around the corner for you and me as well? Studies performed outside the therapeutic realm have added fuel to this fire. In one example, researchers at the University of Washington used scalp electrode recordings (EEG) to control a device called a transcranial magnetic stimulator (TMS), which uses spatially targeted magnetic effects to inactivate brain areas just under the skull. Attaching the EEG and TMS hardware to two separate subjects isolated from each other in different rooms made it possible for the person wearing the EEG to remotely perturb brain activity in the other participant, demonstrating an extremely crude form of the kind of brain-to-brain communication used by such fictional species as *Star Trek*'s Talosians. In another well-publicized case, neuroscientists at the University of California, Berkeley used a computational algorithm to reconstruct imagery from fMRI scans of a subject who was watching a video. The fMRI-based reconstruction looked like a smeared-out version of the original video, inspiring speculation that such methods could be used for a rudimentary form of mind reading. "Like computers, human brains may be vulnerable to hackers," proclaimed a news article reporting the work.

A brand of self-styled technological prophets foretells the extension of today's brain hacking into yet more fantastic innovations that have yet to be realized. "Twenty years from now, we'll have nanobots [that will] go into our brain through the capillaries and basically connect our neocortex to a synthetic neocortex in the cloud providing an extension of our neocortex," predicts the author and engineer Raymond Kurzweil. Kurzweil believes that a consequent merging of human and artificial intelligence will radically change the human condition through a synthesis he and others refer to as *the singularity*. Taking a similar tack, physicist and science popularizer Michio Kaku writes that "one day, scientists might construct an 'Internet of the mind,' or a brain-net, where thoughts and emotions are sent electronically around the world." "Even dreams will be videotaped and then 'brain-mailed' across the Internet," Kaku adds, perhaps evoking the Berkeley imagery reconstruction study. Although many are skeptical of such predictions, conjectures like Kurzweil's and Kaku's garner substantial attention.

The futuristic potential of hacking the brain also influences the US military. For better or worse, the defense establishment's interest goes well beyond the humane goal of rehabilitating wounded soldiers. The Defense Advanced Research Projects Agency (DARPA), which funds some of the military's most cutting-edge projects, aims in part to leverage neuroscience to "optimize human aptitude and performance" on the battlefield. Another major thrust involves "understanding and improving interfaces between the biological and physical world to enable seamless hybrid systems." Dispelling any doubt about what such systems could be used for, a team of DARPA engineers connected a quadriplegic patient named Jan Scheuerman to a BMI much like Cathy Hutchinson's; after demonstrating Scheuerman's ability to guide a robotic arm, the engineers had her take mental control of a simulated F-35, the Defense Department's most advanced warplane. DARPA director Arati Prabhakar presented results from this real-life Wonder Woman at a 2015 conference on the Future of War. "We can now see the future where we can free the brain from the limitations of the human body," Prabhakar announced proudly to the audience.

The idea that hacking the brain will free it from the body's confines reflects much of the fascination with neurotechnology. But it is also an idea that springs from the cerebral mystique and carries with it three fallacies rooted in problems I have discussed throughout this book. First is the concept that brain and body are separable to begin with, a notion that illustrates the extent to which the brain has become a stand-in for the dualist's disembodied soul. We saw in Chapter 5 that this is not only a philosophical error but a biological one, in conflict with the fact that many features of human behavior depend critically on reciprocal interactions between brain and body. The second fallacy lies in the sense that the brain is inherently stronger or less limited than the body. Chapter 2 criticized the discourse that depicts the brain and body as working by different principles, with the brain more abstract and inorganic in its modes of operation. In fact, the biological substrates of brain and body have qualitatively similar weaknesses, such as their limited endurance and capacity, as well as their susceptibility to infection, injury, and decay.

The third fallacy is the view that hacking the brain is a good way to break free of any limitations at all. In practice, no existing device comes close to accomplishing this. Although some of the recent triumphs of human neurotechnology are breathtaking, all are limited to some extent by the physical as well as metaphorical violence of hacking. Even non-invasive brain manipulation with TMS is said to feel like a woodpecker pecking at your head, and the crude telepathy that requires this seems like a poor substitute for good old-fashioned speech. Meanwhile, more meaningful neural interventions require risky brain surgery that few subjects would undergo without dire need. To patients with severe disabilities, the benefits of these technologies are merely restorative, and at most partially so. It is only against a backdrop of devastating dysfunction that a person gaining the ability to control a prosthetic arm or benefit from DBS sounds like a success story, and any healthy teenager with a joystick could fly the simulated F-35 better than DARPA's Wonder Woman. It is certainly worthwhile to keep improving the technology for rehabilitating patients with brain injuries and diseases, but the potential for using such devices to hack additional abilities into or out of healthy brains seems remote, unappetizing, and possibly dangerous. Nevertheless, neurotechnological visions fueled by the cerebral mystique retain a special place in the fantasies of those who think about humanity's evolution as a species, as we shall see.

The US presidential election of 2016 featured a famously strange roster of candidates, but few were more unusual than Zoltan Istvan. As the founder and candidate of the first political party associated with the so-called *transhumanist* movement, Istvan sought to represent "a growing group consisting of futurists, life extensionists, biohackers, technologists, singularitarians, cryonicists, techno-optimists, and many other scientific-minded people" who support the agenda of overcoming death and embracing radical technological change. "Who doesn't want to have their lives made better through science and technology?" asks Istvan. The Transhumanist Party did not officially make it onto any of the state ballots, but Istvan's unlikely bid nevertheless garnered news coverage from mainstream media outlets, an endorsement from Robert F. Kennedy III, and over twenty thousand followers on Twitter.

Istvan is a former journalist whose square jaw and fit physique belie his allegiance to some firmly geeky causes. The aspiring politician made his first media statement with a 2013 novel entitled *The Transhumanist Wager*, which tells the story of a philosopher-king named Jethro Knights, who leads the world into an era of peace, technophilia, and extremely long life spans. Knights, like his real-life philosophical predecessor Immanuel Kant, has a *categorical imperative*—a golden rule that instructs him and his fellow transhumanists to "safeguard [their] own existence above all else." They wager that there is no afterlife and therefore decide that they must do everything they can to achieve immortality. Istvan's hero comes to believe that "to combine brain neurons to the hardwiring of computers in order to download human consciousness [is] the most sensible and important direction for the immortality quest." In the idyllic society he builds, everyone walks around with a computer chip in their head that allows them to communicate in a flash with other people or devices. "To stay youthful, healthy, and competitive with one another," Istvan writes, "people spent money on functionally upgrading their bodies and the efficacy of their brains, and not so much on their wardrobes, cars, and other material possessions." Smartphones and computers were integrated into the brain's neural networks, allowing everyone to remain "always connected, always learning, and always evolving."

Neurotechnologies like the ones Istvan writes of are woven deeply into the fabric of transhumanism, showing how idealization of the brain can shape a far-reaching vision for the future, in which enhancement of individual cognitive abilities appears to be the paramount goal. The transhumanist muse Robert Anton Wilson wrote over thirty years ago of a coming "Intelligence Intensification" that will "expand consciousness and sensitivity to signals and information." Wilson reasoned that "Intelligence

Intensification is attainable, because modern advances in neuroscience are showing us how to alter any imprinted, conditioned or learned reflex that previously restricted us." The translation of neuroscience knowledge into the engineered brain implants and interfaces of the future was foretold in the late 1980s by Fereidoun M. Esfandiary, an Iranian Olympic basketball player who became the world's first self-declared transhumanist and changed his name to FM-2030.

In the imaginations of people like FM-2030, futuristic tools for brain interfacing could include not only souped-up BMIs reminiscent of the *Matrix* movies and *Star Trek*'s Borg, but also the nanometer-scale robots called *nanobots*, which we saw mentioned above. Nanobots would be small enough to swim through the body and communicate with individual brain cells. Although some nanotechnology experts challenge the physical feasibility of recognizable robots at this scale, Ray Kurzweil and others seem to remain staunch believers in the power of such gadgets. Even Nicholas Negroponte, an MIT computer scientist and former director of the school's prestigious Media Lab, once advertised the potential of neural nanobots, explaining that "in theory you could load Shakespeare into your bloodstream and as the little robots get to the various parts of the brain they deposit little pieces of Shakespeare or the little pieces of French if you want to learn how to speak French." Of course, such takeovers of the mind could turn out to be anything but beneficial. A digital video series called *H+* dramatized a plague of injected nanoscale brain interfaces in which a computer virus takes over the implants and incapacitates all of the modified humans who harbor them.

For many transhumanists, as with the fictional Jethro Knights, the road to immortality also runs through the brain. A popularly imagined strategy for achieving indefinite life involves the hypothetical procedure of *uploading* all the content from a person's brain and then downloading it back into a new body or possibly a simulated environment. Here again, the brain functions like a soul, self-contained and separable from the body. "The upload is the posthuman," explains the transhumanist thought leader Natasha Vita-More. "It's the copying and transfer of the brain, your cognitive properties, onto a non-biological system, which could be a computer system. . . . So you would be within a whole different universe of computational matter. And that would be a very beautiful simulated environment." To achieve the uploading, many seem to place their faith in the exhaustive anatomical analysis of brain tissue called connectomics, even though current methods have nowhere near the required throughput to scan an entire human brain, let alone to simulate the biology it represents. To while away the time as technology catches up to their aspirations, some transhumanists therefore turn to cryonics as a means to preserve their brains after bodily death. In Chapter 5, we encountered the Alcor Life Extension Foundation, which offers to freeze clients' brains and store them indefinitely for a fee of $80,000. Alcor is run by Vita-More's husband, Max More, a transhumanist philosopher who plans to have his own brain frozen when he dies. One of Alcor's early clients was the trailblazing FM-2030, who failed to achieve his own dreams of immortality and sadly died of pancreatic cancer at the age of sixty-nine. His frozen head has now been steeping for over fifteen years in a vat of liquid nitrogen at Alcor's headquarters in Scottsdale, Arizona.

The quest for cognitive improvement and immortality through brain technology represents the cerebral mystique and the denial of our biological nature at their most extreme. In its capacity as the gateway to a higher plane of human existence, the brain attains the status of a religious entity. Achieving life enhancements and extensions through neurotechnology implies not only the equation of each person to his or her brain but also a solipsistic focus on manipulating the person's existence by manipulating that one brain alone. This mission largely ignores the interrelated, socially and environmentally dependent texture of human mental life, and trivializes the problems of human society by focusing on individual existential concerns that generally only interest the well-to-do. "Even in cases where social benefits [of transhumanist enhancement] are brought up, these are seen rather as the result of cumulative individualistic interventions," writes ethicist Laura Cabrera. Indeed, it is hard to see places for collective values such as equality, empathy, and altruism in a transhumanist culture where people strive to safeguard their own individual existence—and in particular their brain's existence—above all else.

This agenda is not as far from the mainstream as it might first appear. Although transhumanists may come across as a fringe element, there is plenty of contact between this group and various professional communities. Recent transhumanist conferences have featured leading neuroscientists, and several academic biologists are ostensibly working toward transhumanist aims. Defense organizations like DARPA are heavily influenced by transhumanist ideas about brain technology, as we have seen. In the arena of big business there is likewise evidence that transhumanist goals are gaining traction, as major Silicon Valley entities increasingly invest in anti-aging research. Even outside these power centers, a substantial portion of ordinary people could be attracted to transhumanist promises of living longer and becoming smarter. Some transhumanist ends are after all not so different from the objectives of modern medicine and education, broadly speaking. But how do the movement's proposed *means* for achieving human enhancement compare with more conventional alternatives?

There are many reasons for taking issue with the transhumanists' interest in advancing the human race through purposefully engineering the body, as opposed to letting natural selection run its course. Cultural taboos against "playing God" with the human form abound, but an even more general objection springs from the principle of unintended consequences, which warns us that tinkering in abrupt ways with time-tested physiological processes honed by evolution could go awry. There is the possibility of side effects both in individuals and in groups that have been altered with technologies like brain implants, genetic engineering, or life-extending drugs. A world in which nobody dies could also have severe problems unrelated to the methods by which immortality was achieved. Unless births are severely regulated, power and resources might eventually need to be split among trillions of transhumans, and conflicts would arise in the struggle for Lebensraum. The loss of generational turnover could also deal a great blow to human culture and innovation. The historian of science Thomas Kuhn famously observed that new scientific theories tend to catch on only when recalcitrant "old believers" literally die out. As in science, fresh ideas, fresh ambitions, and fresh faces are desirable in many walks of life. Would we want to jeopardize our potential for creative advancement by crystallizing our society into a fixed state?

Although the transhumanist goal of making everyone smarter seems uncontroversial enough, even this could have downsides, at least from an evolutionary standpoint. Among today's humans there is considerable evidence for a negative correlation between education and fertility. Could it be that transhumans with enhanced cognition will be too busy being brilliant to reproduce? Looking around the planet, one sees quickly that the most abundant and arguably successful organisms are not necessarily those with the greatest intelligence. Beetles, for instance, have been around for approximately one hundred times longer on earth than we have and account for 25 percent of known species, with a global population likely well in excess of ten trillion. The great biologist J. B. S. Haldane once remarked that God seems to have "an inordinate fondness for beetles."

But it is transhumanism's inordinate fondness for brain technology that flavors its peculiar outlook for humankind and reveals the biases that arise from idealizing the brain. While you or I might imagine a future bristling with sleek innovations that surround and empower us, the typical transhumanist wants technology that physically penetrates us, and in particular our heads. It is as if the gadgets we use would be somehow intrinsically stronger or better if they were directly attached to the brain. We could get information from the internet directly into our brains without reading. We could drive our cars simply by thinking, without having to move our hands. We could communicate without having to strain our vocal cords or pump our lungs. In each case, electronic technologies like neural nanobots or brain chips obviate the need for biological components that are thought to do little more than get in the way. In these transhumanist visions of neurotechnology, the brain is carried forward into a digital universe, while the rest of the body is largely left behind.

But why should the technology that benefits our cognition and control directly tap into the brain? The approach seems to be motivated by a desire to get straight to the essence of the person, following the neuroessentialist mantra that "we are our brains." From a practical standpoint, this seems excessively restrictive, however. "Most of the benefits you could imagine achieving through [brain implants] you could achieve by having the same device outside of you and then using your natural interfaces like your eyeballs, which can project a hundred million bits per second straight into your brain," says philosopher Nick Bostrom of

Oxford's Future of Humanity Institute. Suppose, for instance, that you want to give someone the mental impression of a beetle; you have the choice between showing them a photograph and stimulating cells in the brain directly so as to produce the same imagery. Both routes would by definition result in exactly the same pattern of brain activity corresponding to the perception of a beetle, but there is no doubt that simply presenting the picture would be far easier. By contrast, "writing" the beetle percept directly into the brain would require invasive manipulation, as well as a degree of knowledge about brain function that far surpasses what we have at present. Even if sufficient understanding were available, trying to circumvent the biological input and output pathways that surround the brain would require improving on mechanisms that have served humanity well for millions of years.

Familiar real-world examples back up the idea that technology can substantially enhance human mental performance without directly contacting our brains. Over four thousand years ago, the Sumerians of southern Mesopotamia invented an abacus that may have been the world's first computing machine. With devices like this, users could quickly manipulate larger numbers than possible using short-term memory and endogenous thought processes alone. The greatest cognitive aid to humankind was probably the invention of writing systems in societies such as the ancient Near East, Shang dynasty China, and pre-Columbian Mesoamerica. In some sense, the strength of the written message arose precisely from the fact that the brain is out of the equation once its thought content is recorded; this makes written dispatches more reliable than missives remembered in a human messenger's mind. Philosopher Andy Clark argues that external artifacts like abacuses and written records form part of the "extended mind" of the person who uses them, just as purely brain-based processes or functions involving neural implants would. The mind and self are "best regarded as an extended system, a coupling of biological organism and external resources" that need not lie within the skin, Clark wrote in an influential 1998 essay with David Chalmers.

There could be costs to bringing external cognitive resources too close to our brains, even if the logistical barriers against doing so were surmounted. Speaking personally, for instance, I can say that my smartphone has become an indispensible prosthetic aid to my cognition and communication, but one I have no desire to see hardwired into my brain—it's intrusive enough already. Similarly, the computers I use for my scientific research are wonderful number-crunching machines that help me and my colleagues solve our problems in the lab, but would do us no further good if they were embedded in our skulls. If we interfaced such computers directly to our brains, we might find ourselves constantly distracted by them, and our computations on the other end could wind up being disrupted by needless neural input. A different form of cognitive aid I and many Bostonians dream of is something that would make people better drivers, but again the best solution probably lies outside our brains. In this case, industry seems to be congealing around a strategy that is almost completely divorced from human cognition: have the cars drive themselves.

Practices in the therapeutic domain also demonstrate how neurotechnology that works around the brain might be preferable to technology that works within it. In 1968, a swashbuckling Royal Air Force veteran and physiologist named Giles Skey Brindley implanted the visual cortex of a blind patient with an array of eighty brain stimulation electrodes. Passing microcurrents of electricity through the electrodes caused the patient to experience a visual sensation called *phosphenes*—similar to the spots you can sometimes see after rubbing your eyes. The locations of the phosphenes depended on which electrode was stimulated, indicating that a rudimentary form of spatial acuity could be restored by stimulating different electrode combinations. Writing triumphantly of this success, Brindley and his coauthor suggested that the approach would one day enable the blind to "read print or handwriting, perhaps at speeds comparable with those habitual among sighted people." In the ensuing years, however, Brindley became best known not for restoring vision to the blind but for figuring out how to achieve a chemically induced erection; in one instance, he supposedly dropped his pants in front of an audience at a major conference and made his point in the flesh. Meanwhile, Brindley's idea of using brain implants for visual transduction was largely supplanted by the competing strategy of using similar electrode-based arrays farther away from the brain, in the eye itself. Retinal prosthetics often work better not only because of their relative ease of implantation but because they make better use of the body's natural processes for admitting visual information to the brain. Similar advantages have made cochlear implants, rather than auditory cortex brain implants, the dominant therapy for treatable deafness.

Peripheral neurotechnology may also provide an especially promising route for improving human movement and motor function. Researchers have already discovered brain-independent ways to restore movement and control to patients who have lost limbs. Using a technique called *targeted muscle reinnervation*, surgeons can reconnect peripheral nerves from a patient's missing limb to new muscle groups that in turn can control a prosthesis. In 2015, a fifty-nine-year-old man named Les Baugh was given the chance to undergo this procedure at the Johns Hopkins University in Baltimore. Baugh had lost both of his arms in an electrical accident as a teenager. After the neural remapping, he was fitted with two cybernetic arms that mounted over his shoulder stumps and moved on command from his reinnervated chest and shoulder muscles. Baugh was able to learn how to control the limbs after only ten days of training, performing such feats as stacking blocks and drinking from a cup. Unlike "locked-in" patients such as Cathy Hutchinson and Jan Scheuerman, who had lost all ability to communicate with the rest of their bodies, Baugh did not need to control his prostheses via a direct connection to his brain. His brain was barely less involved in controlling his artificial limb movement, however. Motor output from Baugh's brain triggered the reinnervated muscles that directly piloted the limbs, but this process exploited the brain's embodiment in a broader biological milieu, rather than trying to get around it.

To go beyond rehabilitation and enhance the capabilities of healthy individuals, a related approach involves fitting subjects with a so-called *powered exoskeleton*; the exoskeleton gives added strength and rigidity to its wearer via a system of braces and actuators, enhancing the wearer's ability to perform physically demanding tasks. In Marvel's *Iron Man* comics, the fictional Tony Stark controls a powerful exoskeleton using impulses from his brain, but real-life experimental exoskeletons take their input from their wearers' bodies. For instance, the HAL-5 exoskeleton manufactured by the Japanese company Cyberdyne is controlled largely via a set of skin surface electrodes that read impulses from the wearer's musculature and interpret them as movement commands to control the power suit. The suit looks a bit like *Star Wars* stormtrooper armor and enables people of average strength to lift objects as heavy as 150 pounds with little effort.

The closest thing to a superman we could produce today might sport one of Cyberdyne's exoskeletons while enjoying extraordinary communications and computational capabilities afforded by portable or wearable electronic devices. If he saw through walls, it would be because he could pilot a remote-controlled, camera-bearing drone. If he could sense bodies in the dark, it would be because he wore infrared glasses. If he possessed a supercar or superplane, chances are that the vehicle would be super mainly because of its own autonomous control mechanisms, as opposed to a connection to its owner's gray matter. Our modern superman would be a testament to the embodied brain, an individual whose nervous system transduces an extended array of inputs into actions enhanced by peripheral aids, noninvasively interfaced to distributed elements of his natural human physiology. This hero provides a counterpoint to transhumanist visions in which the hacked brain itself is the secret to transcending humankind's limitations. Trying to improve the brain with invasive neurotechnology makes most sense in an imaginary world where the brain is differentiated from its surroundings, solitary, self-sufficient, and soul-like. If one accepts that the brain is a biological organ that functions integrally with the body and environment, then neurotechnological enhancements to human capability no longer need be bound to the brain.

The seeming remoteness of futuristic conceptions of cognitive enhancement does not keep people from worrying about their implications already. In a 2004 essay, the political scientist Francis Fukuyama labeled transhumanism one of "the world's most dangerous ideas" because of the potential threat of transhumanist-style intelligence improvements to conceptions of human equality. "If we start transforming ourselves into something superior, what rights will these enhanced creatures claim, and what rights will they possess when compared to those left behind?" Fukuyama asked. "If some move ahead, can anyone afford not to follow? . . . Add in the implications for citizens of the world's poorest countries—for whom biotechnology's marvels likely will be out of reach—and the threat to the idea of equality becomes even more menacing."

Concerns like Fukuyama's might be more relevant to today's society than we realize. Although the implants and nanobots of transhumanist fantasy may never really come to be, another class of intelligence enhancement is already here. So-called *nootropic* substances—from the Greek words meaning "mind-bending"—are consumable chemicals thought to be capable of improving concentration, memory, and other aspects of cognition. Nootropics are similar in spirit to the magic potion that gave Superman Bill Dunn his powers, albeit not as magical in actuality. The most ubiquitous examples are relatively mild naturally occurring stimulants like nicotine or caffeine, the humble "cognitive enhancer" we considered briefly in Chapter 5. Nootropics also include dietary supplements like omega-3 fatty acids, which are thought to foster positive mood, or the racetams, which may modulate activity of key neurotransmitters in the brain. The most powerful nootropic drugs are well-characterized prescription stimulants like amphetamine and methylphenidate, marketed respectively under the names Adderall and Ritalin as treatments for attention-deficit hyperactivity disorder (ADHD), as well as powerful sleep suppressants like modafinil, which is used in both medical and military contexts to promote alertness.

Although heavy-hitting prescription nootropics in the United States are legal only for approved therapeutic uses, they are widely abused in particular by students seeking an edge in their studies. A common pattern is for students to obtain substances from acquaintances who have legitimate prescriptions, and then to use the drugs nonmedically as aids to binge studying. A 2005 survey of over one hundred four-year colleges in the United States found that an average of 7 percent of students illegally used prescription stimulants, with rates of up to 25 percent at some individual colleges. Several studies have questioned whether prescription-strength nootropics truly boost academic performance, but their prevalence on campuses shows that there is considerable belief in their efficacy. Students who use prescription nootropics illicitly must value the potential rewards offered by these drugs enough to override concern about the possible criminal penalties for being caught.

Nonprescription nootropics are currently perfectly legal, but they are serious business in their own way. Silicon Valley start-ups like Nootrobox and truBrain have raised millions of dollars in investment capital to market products formulated from supposed nootropic ingredients. Merchandise like theirs appears to have a faddish following among the so-called biohackers who want the benefits of prescription cognitive enhancement drugs without the accompanying regulatory hassles. Nootrobox, for example, sells "stacks" of compounded over-the-counter nootropics, ranging from chewable coffee drops to capsules that combine ingredients from the Indian pennywort, the western roseroot, and various vitamins and neurotransmitter analogs. Each ingredient is said to be "generally regarded as safe" by the US Food and Drug Administration, but demonstrations of efficacy are also generally minimal; the company is currently trying to prove the effectiveness of its products through a clinical trial.

Whether you are a struggling student thinking about how to get ahold of prescription study aids or an ambitious entrepreneur wondering whether to spend $100 per month for an extra edge from commercial smart pills, you may already feel some of Fukuyama's questions beginning to apply. If you are competing with nootropic-using peers, can you afford not to follow suit, given the cutthroat nature of the work environment? It is an anxiety that nootropic marketers consciously play on. "If you aren't taking Alpha BRAIN, you are playing at a disadvantage," warns the website of a company called Onnit, pushing its herbal brain-boosting supplement. Many of the competitors themselves agree. Explains businessman and self-help author Tim Ferriss, "Just like an Olympic athlete who's willing to do almost anything, even if it shortens your life by five years, to get a gold medal, you're going to think about what pills and potions you can take." It is exactly this mindset that leads many to dystopian premonitions. "All this may be leading to a kind of society I'm not sure I want to live in," laments *New Yorker* staff writer Margaret Talbot, "a society where we're even more overworked and driven by technology than we already are, and where we have to take drugs to keep up; a society where we give children academic steroids along with their daily vitamins."

Such fears about the encroachment of neurotechnology are the flip side of enthusiasm for neurotechnology's benefits. But both assessments stem from an inflated sense of the brain's significance compared with body and environment, and both may be mistaken in ways that apply also to other inflated concerns in our culture. You might, for instance, have heard the cliché that "one man's terrorist is another man's freedom fighter." The saying highlights the subjectivity with which people tend to use the terrorist moniker to vilify their enemies, plus the fact that there is usually someone around who takes a positive view of whatever nefarious cause a given terrorist espouses. Meanwhile, despite the fact that terrorism rises near the top of voters' concerns in most public opinion polls, *New York Times* columnist Nicholas Kristof points out that in recent years far more Americans have drowned in bathtubs than have been killed by terrorists. Regardless of one's feelings about individual groups, terrorism as a phenomenon seems to get more attention than its impact on society should merit.

Brain technology, with both its advocates and its detractors, fits a similar pattern. We noted the anomalous zeal among transhumanists and others for technological approaches that interact physically with the brain, even though more peripheral technologies can achieve superior results without the danger and complexity associated with direct brain interventions. Similarly, foreboding about the antisocial effects of neurotechnology such as nootropic drugs may reflect an artificial and counterproductive distinction between cognition-enhancing strategies that act within brains as opposed to around them. If one is troubled, as Fukuyama is, by how neurotechnology stands to increase human inequality and promote a hypercompetitive society, one should direct equal concern toward the many activities that influence brains less directly but produce equivalent consequences. To perceive an unnatural threat from technologies that interact with the brain is no more rational than to display unnatural optimism about them. Neurotechnology may be neither a terrorist nor a freedom fighter, but just one of the many facts of life whose influences are complex and context-dependent.

In fact, ethicists who have considered the impact of nootropic use are generally quick to point out the continuum of related phenomena in which "smart drugs" sit. In a 2008 *Nature* magazine commentary on responsible use of cognitive-enhancing drugs, a group of experts led by the director of Stanford University's Neuroscience and Society Program, Henry Greely, draw parallels between nootropic use and improvements like education, nutrition, exercise, and sleep, all of which affect brain function. They argue that "cognitive-enhancing drugs seem morally equivalent to other, more familiar, enhancements" and that "a proper societal response will involve making enhancements available while managing their risks." Another analysis of ethical issues associated with cognitive enhancement commissioned by the British Medical Association makes some of the same comparisons. Putting the debate over artificial cognitive enhancements in context, the BMA panel also emphasizes that "we need to remember that a wide range of social factors directly or indirectly affect health, welfare and social success." "Merely to focus on one, such as an individual's cognitive abilities," they argue, "ignores the fact that many different social determinants affect individuals' ability to thrive physically and psychologically and to succeed socially."

From cradle to grave, the world treats people and their brains very differently. Some babies are born with biological determinants that predispose them to academic achievement, possibly because of differences in the quality of their attention, endurance, memory, or speed. Perhaps more importantly, some babies are also born to pushy parents who give them books instead of toys for their birthdays and start them in afterschool programs from the moment they can talk. Wealth disparities contribute to cognitive enhancement in multiple ways, from parents' capacity to spend time helping their children over educational hurdles to their ability to afford benefits like computers or private lessons. The culture of a household and its broader social context plays an immense role beyond schooling per se, influencing facets of life like emotional well-being, ambition, and health. When kids grow up and leave the family, their childhood engrams stay with them, as most certainly do generic biases introduced by their socioeconomic origins. And make no mistake about it: each socially determined factor affects the brain just as surely as a genetic contribution or a nootropic drug might. The brain is plastic and can be changed by a huge range of inputs. Education and values become imprinted into the brain like other memories, and they influence future behavior as a result. Economic and social security affect stress levels and the body-wide physiological pathways that go with them. For all of these reasons, it seems unlikely that currently available neurotechnologies could significantly worsen what is already an extraordinarily uneven playing field for society's diverse team of eight billion nervous systems.

This is not to say that questions about how to regulate nootropics should not be asked. Particularly given the prevalence of illegal prescription drug use in the United States, as well as limited evidence concerning safety and efficacy of over-the-counter nootropic supplements, some degree of further analysis and regulatory action is certainly called for. But if a goal of governing nootropics and other cognition-enhancing neurotechnologies is to ensure that they do not contribute to the injustices that critics like Fukuyama fear, then it seems at least as worthwhile to think about how to compensate better for the unfair allocation of "soft neurotechnologies," such as pushy parents and competitive communities, that already promote inequality in more powerful ways than any form of drug or device-based brain hacking is likely to.

It is no coincidence that the Titan Prometheus became the patron deity of both hacking and human enhancement, and his story also symbolizes some of the points I have tried to make in this chapter. According to ancient Greek mythology, Prometheus molded the first human from clay and then empowered him by giving him fire stolen from Mount Olympus in violation of Zeus's wishes. As punishment for crossing the king of the gods, Prometheus was condemned to spend eternity chained to a rock, beset by a ravenous eagle that devoured part of his liver every day, until the hero Hercules freed him. "Prometheus stole fire from the gods on behalf of mankind. That's all some youthful hacker outlaws today need know to inspire them to adopt Prometheus as their icon," explains technology commentator Ken Goffman. We can see aspects of the Prometheus legend in the stories of people like Edward Snowden, who shed light on the National Security Agency's inner workings and was thereupon forced into exile, or Aaron Swartz, the hacktivist whose campaign for open access to online resources led to his arrest and subsequent suicide. That Prometheus himself was released from his rock can be seen as a vindication of his inventiveness and an acceptance of his technology. But the creator's release also gives him wider latitude to invent and perhaps be judged again.

I have argued that both hopes and fears about neurotechnology should be unshackled from the brain as Prometheus was from his rock. Because the brain is just a prism through which myriad internal and external influences refract, our aspirations will often be addressed more easily by modifying the influences than by manipulating the prism itself. By uncoupling from the brain our visions for the future of the mind, we vastly expand the scope for new technology development. Similarly, by broadening our concerns about the unwelcome effects of cognition-altering technology from those that work directly on the brain, we might gain renewed motivation to address existing discrepancies in education and culture. Such inequalities affect our brains as definitively as any pill or implant would, and they pervade our society already.

The cerebral mystique constrains our thinking about neurotechnology in much the same way it restricts views about mental illness and the individual's place in society, as discussed in Chapters 7 and 8. In each case, the mystique fosters a tendency to analyze people's problems in terms of their brains alone. To questions about what makes us do what we do, what makes us experience mental pathologies, or what could improve our cognitive abilities, the cerebral mystique offers one answer: the brain. But the fundamental lesson of neuroscience is that the brain is a biotic organ, embedded in a continuum of natural causes and connections that together contribute to our biological minds. This means that the brain cannot be all there is. To any question about altering or explaining human behavior there are actually many answers, at levels embracing not only the brain but the body and environment it resides in. In an age in which self-absorption and self-centeredness have reached epidemic proportions and the socially minded values of previous generations are on the wane, the message that *you are not only your brain* may be one of the most important lessons science has to teach us. Accepting this message involves rejecting myths about the special soul-like qualities of the brain and understanding how the brain is physiologically coupled to its surroundings. It is only by doing so that we can truly grapple with our place as biological beings in a universe of interrelationships.

ten

WHAT IT'S LIKE
TO BE IN A VAT

THIS CHAPTER IS UNLIKE THE OTHERS IN THIS BOOK. IT is the story of how I came to appreciate firsthand the place of my brain in the world. The tale began unexpectedly at a pan-Hispanic tapas restaurant near my office in Cambridge, Massachusetts. My wife Naomi and I had set our sights on La Mente Quebrada since it had opened in the spring, but the place always seemed to be booked weeks in advance. When a solitary time slot opened up on one of our date nights, we leapt at the chance to take it.

As we made our way over from the parking lot, the late October wind whipped through our coats and nipped at our ears. The trees shook in violent fits around us; their balding branches showed that they were fast losing their annual battle against the seasons. My stomach tightened, and the breath sheltered in my lungs, refusing to commune with the unruly air outside.

The establishment we entered offered us refuge from the cold but took its own toll on our ears. Amped-up drumbeats reverberated aggressively off every surface, while the dissonant howling of an apparently deranged singer easily outdid the gusts we had left behind. A young woman near the door made eye contact and mimed a welcome at us with her brightly painted lips.

"We have a reservation," I shouted, doing my best to break through the noise.

We pushed our way past a crowd of buff and bearded hedonists yammering at the bar, and brushed through a beaded curtain. Our surroundings shimmered faintly with gold mosaic tiles and purple velvet upholstery, elements of bordello-inspired decor reflecting the meek output of a motley collection of salvaged chandeliers and sconces. Along with the fixtures, the carcass of a pig was swaying from the ceiling, as if in flight. Mounted above one of the doorways was a trio of large stuffed crows dressed as mariachis; as I stared at them, I nearly walked into a huge bull's head trophy that seemed to lunge at me from one of the walls. We stumbled past these biotic relics and through a noisy thicket of thrumming tables to reach the one empty spot in the establishment, wedged unceremoniously in the back with little or no elbow room.

"So are we going to get the *chapulines*?" asked Naomi.

Although my appetite was meager that night, my curiosity was also aroused by the more exotic items on the menu. The ant egg tacos called out to me, as of course did the sheep's brain omelet. Naomi was particularly spooked by the idea of eating brains, but we decided to give them all a go anyway.

The blow struck just as I had swallowed my first bite of the tortilla— a piercing abdominal pain combined with nausea that sent me careening back toward the charging bull and then on to the toilet. Still bent over the bowl but slowly regaining my bearings, I began to see red. With a shade less strident than the hot hue of our spiced grasshoppers and brighter than the deoxygenated crimson of the sangria we had been sipping, I realized that it was the plain, elemental red that stains the matador's shirt when he has been too slow. It was blood—a lot of it—and it was swirling below me like an invitation to battle.

Naomi had tracked my unstaged departure with a concerned attention, and now I heard her voice straining to call me over the din outside. I washed up as quickly as I could and felt my way feebly out of the bathroom. The dimly lit restaurant seemed even darker than before when I reemerged and made out Naomi's silhouette in front of me. I told her what had happened, and she wanted to bring me to the emergency room right away. My intellect but not my strength consented to be taken. Forging with supreme corporeal effort back through the crowd, noise, and wind, I wanted nothing more than to surrender to a fatal fatigue—to leave my body behind and drift away to some sort of ethereal stasis, where my mind alone could persist in peace.

Our journey took us instead to a hospital admissions department flooded with harsh fluorescent lighting and insistent requests for insurance information. Naomi and I rotated through wards and waiting rooms with a glacial pace that was not in itself unwelcome to me, but at each step of the way sensory intrusions dashed any hope I had for true rest. The rude grip of the blood pressure cuff, the cold slime of ultrasound jelly, and the sharp prick of the phlebotomy needle all added insult to the injury I felt aching incessantly in my stomach. When I was finally installed in a comfortable bed, it was only to have the reptilian cord of an endoscope thrust down my throat. My afflictions were offset by Naomi's silent presence beside me and by the reassurance of her hand on my shoulder. As the tortures tailed off and I slid into sleep, the last thing I perceived was my wife melting into a bleary haze as a chorus of equipment beeped continuously in the background.

I don't know how long I remained unconscious. In my dreams I imagined a succession of further medical persecutions. I was plunged into scanners that whizzed or buzzed unbearably about my body, invaded by surgical instruments that bored into my belly and perforated my stomach, and smothered by heavy metallic drapes that evoked the fate of poor Giles Corey, the accused wizard who was pressed to death during the Salem witch trials. Corey's ordeal had supposedly stretched without respite over two days; I suspect that my troubled slumbers might have been similarly prolonged.

When I awoke, the face I saw was not Naomi's but that of an old man with wire-frame glasses, wispy hair, and a long white beard that matched his lab coat.

"I'm Dr. Peters," he said.

"Where is my wife?"

"She couldn't be here. You and I must meet alone."

We were no longer in the place I had fallen asleep in. Gone was the bustle and beeping of the hospital floor, and instead I was sure I was listening to the soundtrack from a meditation class. The room was spare, and aside from the bed I lay on, the only furniture included a small side table and a chair. There was a closed door in the corner. The walls were blank, save for a large poster across from me. It depicted a magnificent but unnaturally colored mountain range with words superimposed on it: "Never give in; never give in, never, never, never." Winston Churchill, I thought.

"You had stage four metastatic stomach cancer," the old man said. "Your organs failed, and you went into cardiac arrest, but now you're cured."

"Cured?"

"Yes, your brain was saved at your wife's request. It's now being kept alive indefinitely in a life support system. Everything important about you has been preserved, and your body will never again be a problem."

Flummoxed by this news, I didn't know quite how to respond. Of one thing I was certain, however: at least some of my bodily sensations were emphatically still there. The pain and aching of the earlier evening were gone, but I felt nearly constant tingling in my extremities, as if someone was randomly running a feather over my skin. Further proof of my embodiment lay below me on the bed. My arms twitched involuntarily beside me as the feather ran over them, and my legs shifted awkwardly as my attention descended in their direction.

"Your experiences are being simulated by a computer," Peters interjected, joining my train of thought. "Your anatomy has been generated by software."

"This is ridiculous," I exclaimed in disbelief, my patience running short. I would have lost my temper entirely were it not for a lethargy that seemed to hold my emotions in check. My urge to protest persisted, but the anger and anxiety I felt were divorced from the drive to action I would normally have known in such a situation. Even my words were deprived of their usual edge and intonation. I felt that I was speaking with someone else's voice.

"You can't expect me to believe what you're saying. What is going on, and when can I see my wife?"

"Let me explain more to you," the doctor responded, trying to calm me. Suddenly he and the rest of the room faded into nothingness, and I found myself instead in an operating theater. Three enormous surgical lights illuminated the scene. Vital signs monitors hummed and pinged everywhere, and a bank of computers and monitors stood at one side, displaying an array of flat lines. While an anesthesia machine dispensed its soporific vapors, a large wheeled metal tank hissed slightly with some sort of gaseous emission of its own. At the center of it all, a cluster of men and women in scrubs were gathered around a gurney. On the gurney was a human-sized lump completely covered in aquamarine surgical drapes, save for the head. I recognized the face under the patient's anesthesia mask—it was my own.

One of the medics shaved off my hair with an electric razor as I watched, dumbfounded, from my third-person perspective. Was I still asleep? Two of the attendants clamped my head into a vise. Then an authoritative figure stepped up and slit into my scalp with a surgical knife. The real me could almost feel the sting of the blade. "Stop!" I shouted as the surgeon began to pare back my scalp, exposing the skull. The participants froze for a moment and then vanished. I was back in the spartan bedroom, with Dr. Peters again standing before me.

"You were watching a record of the neurosurgery operation you went through," the doctor explained. "Your brain was extracted using the best techniques available at the time. For the past fifty-four years, it has been safely cryoprotected in liquid nitrogen, but we have learned in the meantime how to rejuvenate frozen brains. Now we can connect preserved brains to bioelectronic neural input/output interfaces that simulate realistic experiences. We turned your simulator on today. Congratulations on being recalled to life."

Convinced now that I was still in a dream, I tried to pinch myself with all the force I could muster. Even my most vigorous effort produced no more than the neutral feeling of something innocuous pressing on my skin. I could not tell if it was my strength that was lacking or my sensory faculties. Searching for alternatives, I tried to bite my tongue. Still there was no pain—just the impression of chewing a large piece of gum.

"Your pain responses have been inhibited," my interlocutor told me, again trespassing into my stream of consciousness. "Otherwise loose nerve endings would be causing you constant torment. You won't need pain perception anymore, though; since your body is a simulation, you are immune to injury."

I was both amazed and befuddled by the doctor's insistence on this absurd narrative. Needing all the more to escape the charade, I determined to rely on brute force. I cast off the bedclothes that lay on top of me and ran toward the door. My movements felt unreal, as if I were wafting across the room rather than sprinting. But as I threw myself against the white slab and tried to twist its handle, the barrier seemed unmistakably substantive. Peters made no attempt to stop me as I struggled in vain to flee the chamber.

"This simulated environment does not allow you to leave the room, but you can choose alternative environments and access other features of the simulator using this." He withdrew a small tablet-shaped device from under his lab coat and placed it on the side table. Then he promptly disappeared.

Like a convict in solitary confinement, I alternately raged to myself and questioned my sanity. In the absence of diurnal cycles, these transitions were my only timekeepers. Even they were attenuated by the sense of emotional emasculation I had experienced since first finding myself in the little room. I slept frequently, though I had no basis for quantifying the length of each interlude. I also thought often of my wife and how I might rejoin her. All the while, Winston Churchill's words on the poster in front of me reinforced my resistance to the possibility that Peters's tall tale might actually be true: "Never give in." But I also had moments of weakness when the alternative reality I inhabited beckoned me to accept it.

It was at one such moment that I picked up the doctor's tablet for the first time. Rather than the tablet computers I was used to, this device held only a single button. I pressed it, and the figure of Dr. Peters suddenly reappeared.

"What would you like to do?" he asked.

"I would like to see my wife," I responded without hesitation.

The room dematerialized, and the scene in front of me was suddenly filled with a montage of pictures of Naomi. I recognized the portrait she had on her professional website, as well as pictures from her old photo collections. There were images from our wedding. Interspersed among these were photos I had never seen before—pictures of Naomi looking more lined and perhaps more worn than she had the night we went to the hospital. Several of them depicted Naomi in old age, as she might have appeared in her eighties. Her eyes were a constant, as was her clean-cut, professional appearance. I could see none of the telltale discontinuities or blurs that typically mar edited photographs. These pictures were either real or very skillfully fabricated.

"Where is she now?" I asked Dr. Peters, who had remained as a silent presence in my peripheral vision.

"She passed away eight years ago," he informed me. The montage of pictures dissolved into words, and as I peered at them, I realized I was reading Naomi's obituary. The text advanced on its own as I viewed an account of a life I knew almost as well as my own. But it soon entered a narrative of events I had not seen. I learned that Naomi had become research director of the nonprofit she worked at. She had published a book. She remarried nine years after her first husband died from stomach cancer, but her second husband had died in 2053.

The part of me that had suspended its disbelief of the doctor's narrative was correspondingly more ready to believe this news—that I would never see my wife alive again. But what I knew should be one of the most painful moments in my life was surprisingly neutral. My heart was quiet, and my breathing imperceptible. I felt no lump in my throat and no urge to cry. Although I had always been quick to perspire in moments of stress, my skin remained as dry as my eyes. The most I might have admitted to was a slight increase in the tingling of my arms and legs. I was bothered as much by how unbothered I felt as by the notion of Naomi's demise itself.

Hoping to change the subject, I called on Peters again. "Can you show me where I am now?"

My wife's obituary vanished, and we were now in a large room filled with rows of coal-black counters. The air was moist and pungent with a smell like ripening cheese. Neatly lined up on the benches were numerous transparent cylindrical aquariums, each about a foot in diameter and height. I could recognize the shapes of human brains waving gently like soft pink corals in the refractive fluid of the tanks. Plastic pipes connected the vats to what looked like small refrigerators on the floor below them. Bundles of tubes and fibers emerged also from the brains themselves, terminating in connectors mounted in the walls of each chamber. On the outside, these connectors in turn were attached to a jumble of equipment that hummed and pulsated on shelves arrayed above the aquariums. Wires were everywhere, snaking around the machines and cascading in thick masses across overhead trays that traversed the room.

"This brain is you," said Dr. Peters as we approached a vat labeled 2017-13.

The organ in tank 2017-13 seemed perfectly unexceptional, as far as I could make out. It was just one of the scores of similar trophies in this laboratory, each of which must once have been part of a complete human being with his or her unique situations, stories, and struggles. As I gazed at the features bulging at me through the curved glass of the container, I had the impression of a mass of uncooked sausage wound up into a large, untidy knot. Close up, I could see a filamentous web of purple vessels that spread like a dark mold over its surface. The thin cables that connected the brain to its interface emerged like flesh-eating worms from inner depths of the tissue. Were all my interests, passions, hopes, and talents shrunk to this little measure? Even in my emotionally blunted state, I was revolted by the idea that my identity had been reduced to this moribund thing.

I reached out and touched the tank. It was warm, and it vibrated faintly with a regular double beat that presumably arose from the support equipment connected to it. As my thoughts became entrained to the gentle rhythm, I felt a sudden urge to respond with a violent shove. But when I pushed against the tank, it wouldn't budge. With my other hand I grabbed at the connections emerging from the aquarium and yanked hard on them. Again there was no result; the cables did not even flex.

Dr. Peters raised his voice from somewhere near my right shoulder. "You are not physically in this room. Although you can experience it through your simulator, there is nothing you can do to change it."

Once more, I discovered that there was no exit from my improbable imprisonment.

As I gradually grew accustomed to the boundaries of my new reality, I also began to discover and partake of its freedoms. I could summon Dr. Peters at the touch of my button and ask to be shown anything or taken anywhere I wished. With the doctor as my guide, I went to places I'd always longed to see and learned about things I'd always wanted to know. I watched the sun set over the Dalai Lama's palace in Lhasa, and I visited the tomb of the great conqueror Tamerlane in Samarkand; I climbed the Bandiagara escarpment in Mali and visited the great mud mosque at Djenne; I scoured the ruins of Roman Palmyra before their tragic destruction; I swam with the last of the blue whales and walked on the surface of Mars.

I learned that I had access to almost every book or article ever written, every movie ever shown in theaters, every program ever aired on TV or radio, as well as a vast collection of recorded performances, exhibitions, and lectures. Surfing through simulator space, I bounced from one activity to the next, guided only by the compass of free association. I lost count of how many times I saw *The Godfather Part IX*, in both its 4-D and 5-D multisensory releases. Meanwhile, I repeatedly attended the 2043 Bayreuth opera festival without purchasing a single ticket, and I developed an appreciation for Patuphony, a mid-twenty-first-century art form that blends classical string improvisation with Oceanic martial arts. I was uplifted by the utopian internationalism of Mina al-Mahzuz's speech at her inauguration as the fifty-fifth president of the United States. And I sat in on the descendent of a bioengineering course I myself used to teach at MIT—now run by a professor born the year I died. But most edifying was the seminar I took on neural simulation technology.

Through this class I learned about the advances in bioelectronic technology and understanding of sensorimotor neurophysiology that made whole-brain interfacing possible, and that seemed to underlie my current experiences as a brain in a vat.

Although my tastes were those of an adult, my life was a child's. The eons I frittered away on whatever whim caught me were spent without concern for time wasted or things left undone. There was nothing to interrupt my interminable succession of self-indulgent explorations. Even the trivial burdens of juvenility were absent. I had nobody to wake me up in the morning, tell me to brush my teeth, or call me to meals. The calls of nature were also silenced, and I had few biological urges. I never needed to change my clothes, wash, or comb my hair; in simulator space I was always well-groomed. I did feel tired occasionally. On those occasions, I generally asked Dr. Peters to send me back to my little room to sleep, but I later learned that this step was not necessary. I once fell asleep while rafting on the whitewater rapids of the Colorado River; instead of drowning or being dashed to pieces on the rocks, I awoke once more in the familiar room, with Churchill's steely injunction there as usual to welcome me.

I never learned why this quotation had wound up on the wall of my home room. Perhaps it had been put there for no reason in particular, like the posters in a dentist's sitting area, but my urge was still to ascribe meaning to it. That meaning had shifted over time, however. By now, the phrase "never give in" had ceased to represent my resistance to Dr. Peters. Instead it seemed to instruct me to embrace the strange immortality I had achieved. I had finally accepted the account of my near death from stomach cancer, the preservation of my brain, and the neurological afterlife I had been plunged into. "Never, never, never" expressed my nervous system's uncompromising defiance of nature's ultimate defeat and its attainment of an apparently eternal postcorporeal existence.

But these words also kept me aware of the other "nevers" that qualified my existence. I was destined never to see my wife, my family, my friends, or for that matter any living soul, save in the highly constrained forms the simulator gave me access to—pictures, videos, or voyeuristic clips of virtual reality. I would also never again experience the physicality of my former life. Although the neural interface supplied much of the sensory information I could expect from my simulated activities, there were notable gaps. Food, for instance, had entirely lost its significance to me; I felt no hunger and savored no sustenance. Sports were reduced to the status of arcade games, physically effortless but devoid of hormonal highs. Even my greatest adventures never offered rewards I would have known in real life. I could climb Everest without the slightest exertion, fearing no danger, but experiencing no sense of achievement as I surmounted its summit. Despite the trappings of my simulated form, I had no muscles to tense as I sought narrow footholds on the rock and ice, no breath to catch when the wind whipped around me, and no heart to pound harder if I faltered in unknown terrain. If I surrendered to fatigue on the slopes, it was not because of exhaustion but because of boredom.

As I continued my journey as a brain in a vat, boredom also replaced yearning for things I had left behind. Ennui was exacerbated by the emotional numbness I had felt from my first moments with Dr. Peters—a handicap I now understood was due to the absence of brain-body interactions my neural interface could not emulate. The result was that I was able to view the most fabulous natural vistas or the most horrific scenes of human suffering with near equanimity. In the absence of emotional investment, I wearied of the perpetual acquisition of knowledge and experience in simulator space. There was nothing to apply it toward and nobody to share it with. Without the ability to interact with the real world, without tasks worthy of taking on, without even the simple challenges of everyday corporeal existence, nothing in my life had a discernable purpose. I was merely a receptacle for input from the neural interface—as much a prosthetic attachment to it as it was to my little gray cells.

I longed to be reconnected to reality, to have my brain put back into an actual body, whether or not it was my own. Even in the person of the humblest pauper or the most destitute drug addict my nervous system could find redemption. The pain of poverty or disease would be offset by the chance to strive toward goals and develop commitments that might bring genuine gratification to myself or others around me. In any whole person and any social context, my brain would be more at home than in the degraded position it now occupied. It could be wiped clear of its memories and reborn in the head of an infant, but it would still have the chance to be reignited with the ardor and ambition that once lent value to my embodied self. Given what I assumed was the impossibility of attaining this, I started to wish instead for the end I had miraculously escaped fifty-four years earlier. Perhaps my incubator would fail, or my tissue would suddenly succumb to an infection. I had literally forever to wait.

But instead my brain continued to rush with increasing rapidity from one pursuit to the next, its attention span growing ever shorter. What my hemispheres demanded the simulator supplied, and each vignette in turn triggered a further stop along the way to nowhere. The device and I would dash together from a class on quantum gravity, to a visit to the city of Einstein's birth, to an exploration of medieval trade guilds, to a reenactment of the Diet of Worms. Or we would dart from a virtual re-creation of Darwin's voyages, to studying speciation in Madagascar, to following the Austronesian migrations, to surfing the seas off Southern California. Without clear perception of time, it was unclear to me whether a voyage around the world took eighty days, hours, minutes, or years. At any moment, where we ended up might depend on a glimpse of text, the glance of a face, or a glint of light offered by the simulator, each of which had the power to evoke the next move almost reflexively from my tethered brain. My thoughts were thrown into dizzying disarray, and any sense of direction vanished into an undifferentiated vortex of constant variation. Until everything changed again.

On that day my simulator died. Or it may have been at night—in my current state I had no way of knowing. Nothing went black. The blur of nervous stimulation I had been caught in merely melted into a mosaic of shifting colored spots that blossomed and faded like fireworks in the distance. The tingling in my phantom limbs became more stochastic, rising and falling unpredictably to the tones of a chaotic ringing that had taken over my auditory system. At one moment, I felt a frisson travel up my now invisible body from toes to face, faintly caressing me. At the next, a bright flash appeared to my right and then quickly slipped away into a swarm of scurrying dots. I noticed a subtle rhythm, an andante of sharpening and flattening notes barely perceptible against the rest of the cacophony, pulsing with a double beat like that of the mechanical system that kept my brain alive. Inexplicable gaps intruded into my awareness; perhaps they were fits of narcolepsy or spasmodic activity originating somewhere in my cerebrum. Short dreams sometimes congealed out of the entropy, starring familiar people and places. Once I saw my wife again at La Mente Quebrada, but then a massive bull rushed out of the shadows at me, and everything disappeared into a complex of red and purple blotches.

In my last moments of sentience I struggled to dictate this book to whatever remained of my neural interface. But without interpretable input to reinforce my impressions, I gradually lost the ability to compose my thoughts or distinguish true from false memories. Eventually every engram became suspect, and nothing remained to anchor me. As coherent imagery grew scarcer, my senses themselves began to blend into one another. I could not tell anymore whether a phantom sound was heard and not seen, or whether a phantom touch was felt and not tasted. The complexity of my thoughts collapsed, and the language of my mind itself was transfigured. Words and pictures ceased to be the building blocks of my ideas and gave way to elemental sensations of varying frequency, duration, and intensity that played to me like an orchestra of mystical instruments. With the simulator gone, these feelings must have arisen from nothing more than the machinery that maintained my remains, ripples in the fluid of the vat, drifts in the conditions of the room in which it resided, or the warmth of an occasional passing stranger. Any of these stimuli could perturb the delicate equilibrium of cells and chemicals that made up my surviving organ, triggering a cascade of responses that sometimes rose into consciousness. My identity dissolved into the milieu.

I had been wrong to recoil from the sight of my brain in its chamber; it had never been anywhere else. The astounding transitions I had undergone since my excerebration were merely the result of substituting simpler brain containers for the more complex, organic one I had started with. Whether ensconced into a human form, interfaced to the simulator, or just steeping passively on life support, my brain itself was always doing the same thing—accepting input from its surroundings and transducing it into actions in the world, output to the neural interface, or whatever subtle emissions might be released into the waters that bathed it. Any sensations I felt or thinking I did were just steps along the way. The simulator failed to give me the experience of being complete because it lacked sufficient functionality to replicate the biological roles of my deceased body and the complexity of the rich context I had enjoyed in corporeal form.

But even if the simulation were improved, the brain in the vat would never be me. I was the brain, and the vat, and the room, and the world around it. I was my story, and my society, and the simulations, and every stimulus that affected me. The organ that contained my memories embedded in its jumble of cellular wiring and neurochemical stew was a special part of me, but only a part contiguous with the whole. Much of what made me myself arose because of what the environment did to my brain, rather than what my brain accomplished on its own. I witnessed this from the upheavals I underwent as my body was obliterated and the input simulator came to control my activities. Even in my earlier embodied state, the person I was and the things I did had been as much a product of the embrace between my physiology and my surroundings as they were in the inanimate container. On the night of my surgery, it was somatic signals, sensory cues, and social interactions that triggered my decision to go to the restaurant, determined the onset of my sickness, sent me to the hospital, and governed my experience of the ordeal. Although events could have unfolded otherwise if I had possessed a different brain that evening, they might well have been more similar than dissimilar.

Dr. Peters had preached the message of the cerebral mystique—the view that everything important about me lay in my brain. He promised that my body would no longer be a problem, and he defined me as an isolated mass of neural tissue floating in an aquarium. Under his guidance, my brain entered its afterlife as a soul might enter heaven. But in designing my neural paradise, Peters and his programmers had drawn a false distinction between brain and body, and between brain and environment. Their simulation ignored the most fundamental lesson of neuroscience: that our brains are biotic entities woven organically into a physical world from which they cannot be extricated without grave loss. When the world

I had known was stripped from my nervous system, I became a fractional person, and the life I was recalled to remained radically incomplete.

We humans have struggled for millennia to define the essence of ourselves as individuals. The ancient Egyptians believed in a tripartite soul composed of the *ka*, the *ba*, and the *akh*—entities that separately encapsulate the properties of being alive and having a unique personality. The oldest Indian writings describe *atman*, a life principle that transmigrates from one being to another over repeated cycles of birth, death, and rebirth. The Pentateuch gave us *nefesh*, an ephemeral spirit that dies with its owner, while classical European culture holds that we each possess an immortal soul, identified in the New Testament by the Greek word *psyche*. Today many people are coming to think that we are our brains, compartmentalized reservoirs of vast complexity that employ mysterious means to direct us through our lives. My book has focused largely on the scientific and practical limitations of this new creed.

The brain is special because it helps orchestrate our behavior *without* distilling us down to an essence. It is a transit point for myriad influences that work jointly on us and through us. In an enlightened age where the brain's function as biological mediator of manifold factors is appreciated, we should have greater ability to look both within and beyond the individual for sources of virtue, intellect, success, and pathology. We should be able to formulate better solutions to many of our challenges at home and in society, with medicine and technology, and also with justice. We should gain insight into how the circumstances of others might have acted on our brains had we been in their places, and as a result we would more easily comprehend the tribulations of less fortunate people. The more we understand this, the more we will understand each other, and the faster we will advance together.

ACKNOWLEDGMENTS

I AM GRATEFUL TO MANY PEOPLE FOR HELPING ME COMplete this book and see it into print. I owe a particular debt to my late colleague Suzanne Corkin, who inspired me with her own book, advised me early in my writing process, and connected me into the publishing business. Sue was undoubtedly a catalyst without whom my project would not have gotten off the ground, and I cannot thank her enough. I am also particularly grateful to Nancy Kanwisher for advice and encouragement at the earliest stages of the work, and for further insightful interactions along the way.

Several friends and colleagues gave me prized input at later stages. Comments from Robert Ajemian, Aviad Hai, Charles Jennings, and Laura Schultz all helped tune the work as it advanced and matured. Detailed notes from Aviad and Charles were particularly crucial. Informal discussions with many others also contributed, and I would especially like to thank my colleagues at MIT, my students and tutees, and members of my laboratory who knowingly or unknowingly participated in such conversations.

I am tremendously grateful to my agents, Kristina Moore and Andrew Wylie, who saw potential in my project and took me on. Kristina has been a wonderful advocate, finding me a publisher and helping me negotiate a very different set of professional ropes from the ones I am used to. The community at Basic Books has been an essential asset. I give special thanks to my editors TJ Kelleher and Hélène Barthélemy for their extensive and thoughtful criticism, and for their enthusiasm about the book. I am also much obliged to Carrie Napolitano, Collin Tracy, Beth Wright, Connie Capone, and Kelsey Odorczyk for their contributions.

Most of all, however, I am indebted to my family. I was fortunate to have been born into a microcosm of academia, and I doubled this pleasure by marrying into another one. My parents, Jay and Sheila Jasanoff, set me along the path I took. Although I betrayed their humanist leanings by becoming a scientist, writing this book has provided an opportunity for detente. My mother was able to give my entire manuscript a close and informed read, and her editing was invaluable. My sister Maya Jasanoff helped coach me through various stages of the publishing process, and also provided much appreciated input on some of the text. My uncle-in-law Boris Katz gave me useful feedback about parts of the book as well. My mother-in-law Anya Shpirt, her partner Pavel Zaslavsky, my father-in-law Victor Kac, and his wife Lena Budrene also helped in indirect but important ways.

The book is dedicated to the two most important people in my life, my wife Luba and daughter Nina. Both of them have been very patient with the project, which used up a substantial portion of what would otherwise have been family time. Luba gave consistent encouragement and support for the book (except for the last chapter), and became my primary editor and a sounding board for key ideas. Nina is disgusted by anything brain-related, but is still the apple of my eye.

NOTES

NOTES TO THE INTRODUCTION

2 **A famous philosophical thought experiment:** Hilary Putnam, *Reason, Truth, and History* (New York: Cambridge University Press, 1981).

2 **At just twenty-three:** Amy Harmon, "A Dying Young Woman's Hope in Cryonics and a Future," *New York Times*, September 12, 2015.

2 **Others have taken Suozzi's path:** Alcor Life Extension Foundation, the company that froze and preserved Kim Suozzi's brain, lists on its website dozens of other clients who have had their brains or heads salvaged and maintained in a similar manner to Suozzi.

3 **"Men ought to know that":** Hippocrates of Kos, quoted in Stanley Finger, *Minds Behind the Brain: A History of the Pioneers and Their Discoveries* (New York: Oxford University Press, 2000).

NOTES TO CHAPTER 1

11 **My next encounter:** "Introduction to Neuroanatomy," Massachusetts Institute of Technology, 2001.

12 **Animals have had brains:** L. L. Moroz, "On the independent origins of complex brains and neurons," *Brain, Behavior and Evolution* 74 (2009): 177–190.

12 **over 80 percent:** S. Kumar and S. B. Hedges, "A molecular timescale for vertebrate evolution," *Nature* 392 (1998): 917–920.

12 **The human brain has an elastic:** N. D. Leipzig and M. S. Shoichet, "The effect of substrate stiffness on adult neural stem cell behavior," *Biomaterials* 30 (2009): 6867–6878.

12 **that of Jell-O:** Jennifer Hay, *Complex Shear Modulus of Commercial Gelatin by Instrumented Indentation*, Agilent Technologies, 2011.

12 **A typical brain is roughly:** Henry McIlwain and Herman S. Bachelard, *Biochemistry and the Central Nervous System*, 5th ed. (Edinburgh, UK: Churchill Livingstone, 1985).

12 **A quarter pound of beef brain:** "National Nutrient Database for Standard Reference Release 28, Entry for Raw Beef Brain," US Department of Agriculture, March 18, 2017.

13 **Archeological findings:** J. V. Ferraro et al., "Earliest archaeological evidence of persistent hominin carnivory," *PLoS One* 8 (2013): e62174.

13 **In evolutionary terms:** Craig B. Stanford and Henry T. Bunn, eds., *Meat-Eating and Human Evolution* (New York: Oxford University Press, 2001).

13 **Nonhuman carnivorous families:** L. Werdelin and M. E. Lewis, "Temporal change in functional richness and evenness in the eastern African Plio-Pleistocene carnivoran guild," *PLoS One* 8 (2013): e57944.

14 **Although other carnivores:** Ferraro et al., "Earliest archaeological evidence."

14 **Celebrity chef Mario Batali:** Mario Batali, "Calves Brain Ravioli with Oxtail Ragu by Grandma Leonetta Batali," www.mariobatali.com/recipes/calves-brain-ravioli/ (accessed March 18, 2017).

14 **Traditional forms of the hearty:** Diana Kennedy, *The Cuisines of Mexico* (New York: William Morrow Cookbooks, 1989).

14 **Truly festive brain:** In contrast to Christians and Jews, Muslims believe that Abraham offered Ishmael rather than Isaac as sacrifice.

14 **Although it is a myth:** Ian Crofton, *A Curious History of Food and Drink* (New York: Quercus, 2014).

15 **Kuru is a disease:** P. P. Liberski et al., "Kuru: Genes, cannibals and neuropathology," *Journal of Neuropathology & Experimental Neurology* 71 (2012): 92–103.

15 **"To see whole groups":** D. C. Gajdusek, *Correspondence on the Discovery and Original Investigations on Kuru: Smadel-Gajdusek Correspondence, 1955–1958* (Bethesda, MD: National Institute of Neurological and Communicative Disorders and Stroke, National Institutes of Health, 1975).

15 **"squeezed into a pulp":** Shirley Lindenbaum, *Kuru Sorcery: Disease and Danger in the New Guinea Highlands*, 2nd ed. (New York: Routledge, 2013).

16 **Even in the sixth century:** Dimitra Karamanides, *Pythagoras: Pioneering Mathematician and Musical Theorist of Ancient Greece*, Library of Greek Philosophers (New York: Rosen Central, 2006).

16 **Consumption of offal:** Nina Edwards, *Offal: A Global History* (London: Reaktion Books, 2013).

16 **A search of a popular online:** Search in www.allrecipes.com for recipes containing liver, stomach, tongue, kidney (and not bean), or brain, performed March 4, 2014.

16 **A 1990 study of food preferences:** Katherine Simons, *Food Preference and Compliance with Dietary Advice Among Patients of a General Practice* (PhD thesis, University of Exeter, 1990).

16 **The participants' tendency:** S. Mennell, "Food and the quantum theory of taboo," *Etnofoor* 4 (1991): 63–77.

16 **Some of these cells:** S. M. Sternson and D. Atasoy, "Agouti-related protein neuron circuits that regulate appetite," *Neuroendocrinology* 100 (2014): 95–102.

17 **A remarkable 1945 study:** J. B. Ancel Keys, Austin Henschel, Olaf Mickelsen, and Henry L. Taylor, *The Biology of Human Starvation* (Minneapolis: University of Minnesota Press, 1950).

17 **"Hunger made the men":** D. Baker and N. Keramidas, "The psychology of hunger," *Monitor on Psychology* 44 (2013): 66.

17 **Phrenology's founder, Franz Gall:** Stanley Finger, *Minds Behind the Brain: A History of the Pioneers and Their Discoveries* (New York: Oxford University Press, 2000).

18 **The book sold over:** P. Wright, "George Combe—phrenologist, philosopher, psychologist (1788–1858)," *Cortex* 41 (2005): 447–451.

18 **Noteworthies from Abraham Lincoln:** William Douglas Woody and Wayne Viney, *A History of Psychology: The Emergence of Science and Applications*, 6th ed. (New York: Routledge, 2017).

18 **Brains of some of Europe's:** Stephen J. Gould, *The Mismeasure of Man* (New York: W. W. Norton, 1996).

19 **In a particularly colorful:** Brian Burrell, *Postcards from the Brain Museum: The Improbable Search for Meaning in the Matter of Famous Minds* (New York: Broadway Books, 2004).

19 **By a curious quirk of fate:** R. Schweizer, A. Wittmann, and J. Frahm, "A rare anatomical variation newly identifies the brains of C. F. Gauss and C. H. Fuchs in a collection at the University of Göttingen," *Brain* 137 (2014): e269.

19 **Wagner noted that Gauss's sulci:** Gould, *The Mismeasure of Man*.

19 **We now know:** M. D. Gregory et al., "Regional variations in brain gyrification are associated with general cognitive ability in humans," *Current Biology* 26 (2016): 1301–1305.

19 **The largest brain collection:** Harvard Brain Tissue Resource Center, McLean Hospital, Harvard University, hbtrc.mclean.harvard.edu (accessed March 21, 2017).

19 **"enhance public awareness":** George H. W. Bush, "Presidential Proclamation 6158," 1990.

20 **"Blueprint for Neuroscience Research,":** R. F. Robert, W. Baughman, M. Guzman, and M. F. Huerta, "The National Institutes of Health Blueprint for Neuroscience Research," *Journal of Neuroscience* 26 (2006): 10329–10331.

20 **In 2013, both the US federal:** Office of the Press Secretary, "Fact Sheet: BRAIN Initiative," The White House, 2013; HBP-PS Consortium, *The Human Brain Project: A Report to the European Commission*, 2012.

20 **Ever-increasing participation:** "Annual Meeting Attendance (1971–2014)," Society for Neuroscience, www.sfn.org/Annual-Meeting/Past-and-Future -Annual-Meetings/Annual-Meeting-Attendance-Statistics/AM-Attendance -Totals-All-Years (accessed March 21, 2017).

20 **The number of print books:** G. E. Moore, "Cramming more components onto integrated circuits," *Proceedings of the Institute of Electrical and Electronics Engineers* 86 (1965): 82–85.

20 **Of the 5,070 "brain" books:** Search on www.amazon.com for "science and math" books with key word "brain" (print books only), performed May 2014.

20 **Over the same time span:** Search on www.pubmed.com for US National Library of Medicine records with key word "brain" or "neuron," performed May 2014.

20 **Psychology is reported:** Carly Stockwell, "Same As It Ever Was: Top 10 Most Popular College Majors," *USA Today*, October 26, 2014.

20 **The number of students graduating:** "Table 322.10: Bachelor's Degrees Conferred by Postsecondary Institutions, by Field of Study: Selected Years, 1970–71 Through 2014–15," National Center for Education Statistics, nces .ed.gov (accessed March 22, 2017).

20 **As a child:** Karen W. Arenson, "Lining Up to Get a Lecture: A Class with 1,600 Students and One Popular Teacher," *New York Times*, November 17, 2000.

22 **The brain I remember best:** "The Brain of Morbius," *Doctor Who*, season 13, episodes 1–4, directed by Christopher Barry, British Broadcasting Corporation, January 3–24, 1976.

22 **Even on the covers of neurology textbooks:** Eric R. Kandel, James H. Schwartz, and Thomas M. Jessell, eds., *Principles of Neural Science*, 3rd ed. (New York: Appleton & Lange, 1991); Mark F. Bear, Barry W. Connors, and Michael A. Paradiso, *Neuroscience: Exploring the Brain*, 3rd ed. (Philadelphia: Lippincott Williams and Wilkins, 2006); David E. Presti, *Foundational Concepts in Neuroscience: A Brain-Mind Odyssey* (New York: W. W. Norton, 2015); Paul A. Young, Paul H. Young, and Daniel L. Tolbert, *Basic Clinical Neuroscience*, 3rd ed. (Philadelphia: Wolters Kluwer, 2015).

23 **"It must be painful":** Arianna Huffington, "Picasso: Creator and Destroyer," *Atlantic* (June 1988).

23 **Some images of glowing brains:** C. G. Jung, *Wandlungen und Symbole der Libido* (Vienna: Franz Deuticke, 1912).

24 **The "feminine mystique":** Betty Friedan, *The Feminine Mystique* (New York: W. W. Norton, 1963).

24 **This cultural movement:** Edward Said, *Orientalism* (New York: Pantheon Books, 1978).

25 **Freud wrote that transference:** Sigmund Freud, *An Autobiographical Study*, translated and edited by James Strachey, *Complete Psychological Works of Sigmund Freud* (New York: W. W. Norton, 1989).

NOTES TO CHAPTER 2

28 **Galen's place in the pantheon:** Penny Bailey, "Translating Galen," Wellcome Trust Blog, blog.wellcome.ac.uk/2009/08/18/translating-galen, August 18, 2009.

28 **Although Hippocrates of Kos:** Stanley Finger, *Origins of Neuroscience: A History of Explorations into Brain Function* (New York: Oxford University Press, 2001).

28 **Galen's vote for the brain:** Gladiator games are a thing of the past, but other forms of brain injury continue to inform neurologists and neuroscientists and have led to important discoveries in modern times as well. Examples are discussed elsewhere in the book.

28 **One famous experiment:** C. G. Gross, "Galen and the squealing pig," *Neuroscientist* 4 (1998): 216–221.

29 **The *rete* figured prominently:** Edwin Clarke and Kenneth Dewhurst, *An Illustrated History of Brain Function: Imaging the Brain from Antiquity to the Present* (San Francisco: Norman Publishing, 1996).

29 **"quite fail to produce":** Andreas Vesalius, *De Humani Corporis Fabrica*, quoted in Charles J. Singer, *Vesalius on the Human Brain: Introduction, Translation of Text, Translation of Descriptions of Figures, Notes to the Translations, Figures* (London: Oxford University Press, 1952).

29 **"Or is it fortune's work":** *The Poetical Works of John Dryden*, edited by W. D. Christie (New York: Macmillan, 1897).

29 **Long before Dryden:** Plato, *Phaedrus*, translated by C. J. Rowe (New York: Penguin Classics, 2005).

30 **"an enchanted loom":** Charles S. Sherrington, *Man on His Nature* (Cambridge, UK: Cambridge University Press, 1940).

30 **Sherrington's fibrous motif:** K. L. Kirkland, "High-tech brains: A history of technology-based analogies and models of nerve and brain function," *Perspectives in Biology and Medicine* 45 (2002): 212–223.

30 **In his book *The Engines of the Human Body*:** Arthur Keith, *The Engines of the Human Body: Being the Substance of Christmas Lectures Given at the Royal Institution of Great Britain, Christmas, 1916–1917* (London: Williams and Norgate, 1920).

30 **Critics have objected:** J. R. Searle, "Minds, brains, and programs," *Behavioral and Brain Sciences* 3 (1980): 417–457; R. Penrose, *The Emperor's New Mind: Concerning Computers, Minds, and the Laws of Physics* (New York: Oxford University Press, 1989).

30 **One of the most memorable episodes:** "Spock's Brain," *Star Trek*, season 3, episode 1, directed by Marc Daniels, CBS Television, September 20, 1968.

30 **The robots of science fiction:** Isaac Asimov, *I, Robot* (New York: Gnome Press, 1950); *The Hitchhiker's Guide to the Galaxy*, directed by Garth Jennings (Buena Vista Pictures, 2005).

31 **In contrast, many of the real-life robots:** M. Raibert, K. Blankespoor, G. Nelson, R. Playter, and the BigDog Team, "BigDog, the rough-terrain quadruped robot," *Proceedings of the 17th World Congress of the International Federation of Automatic Control* (2008): 10822–10825; S. Colombano, F. Kirchner, D. Spenneberg, and J. Hanratty, "Exploration of planetary terrains with a legged robot as a scout adjunct to a rover," *Space 2004 Conference and Exhibit, American Institute of Aeronautics and Astronautics* (2004): 1–9.

31 **Von Neumann argued:** John von Neumann, *The Computer and the Brain* (New Haven, CT: Yale University Press, 1958).

33 **Action potentials spread spatially:** R. D. Fields, "A new mechanism of nervous system plasticity: Activity-dependent myelination," *Nature Reviews Neuroscience* 16 (2015): 756–767; Mark Carwardine, *Natural History Museum Book of Animal Records* (Richmond Hill, ON: Firefly Books, 2013).

33 **Most neurons fire action potentials:** A. Roxin, N. Brunel, D. Hansel, G. Mongillo, and C. van Vreeswijk, "On the distribution of firing rates in networks of cortical neurons," *Journal of Neuroscience* 31 (2011): 16217–16226.

33 **The human brain contains:** Intel Skylake processors included in 2016 Apple Macbook Pro laptops, for instance, contain close to two billion transistors, about fifty times fewer than the number of neurons in a human brain.

34 **Scientists have found a group:** E. Aksay et al., "Functional dissection of circuitry in a neural integrator," *Nature Neuroscience* 10 (2007): 494–504.

34 **To make this possible:** A. Borst and M. Helmstaedter, "Common circuit design in fly and mammalian motion vision," *Nature Neuroscience* 18 (2015): 1067–1076.

35 **Schultz's group studied a task:** W. Schultz, "Neuronal reward and decision signals: From theories to data," *Physiological Reviews* 95 (2015): 853–951.

35 **Remarkably, the behavior of dopamine:** Richard S. Sutton and Andrew G. Barto, *Reinforcement Learning: An Introduction* (Cambridge, MA: MIT Press, 1998).

35 **In a further parallel:** Claude E. Shannon and Warren Weaver, *The Mathematical Theory of Communication* (Urbana: University of Illinois Press, 1998).

36 **The mathematical formalisms:** Fred Rieke, David Warland, Rob de Ruyter van Steveninck, and William Bialek, *Spikes: Exploring the Neural Code*, (Cambridge, MA: MIT Press, 1997).

36 **In a 2010 book:** C. R. Gallistel and Adam Philip King, *Memory and the Computational Brain: Why Cognitive Science Will Transform Neuroscience* (Hoboken, NJ: Wiley-Blackwell, 2010).

36 **The Turing machine processes:** A. M. Turing, "On computable numbers, with an application to the Entscheidungsproblem," *Proceedings of the London Mathematical Society* s2–42 (1937): 230–265.

36 **The authors thus challenge:** S. Tonegawa, X. Liu, S. Ramirez, and R. Redondo, "Memory engram cells have come of age," *Neuron* 87 (2015): 918–931.

36 **John von Neumann's own early efforts:** Norman Macrae, *John von Neumann* (New York: Pantheon Books, 1992).

37 **With the wave equation:** Erwin Schrödinger, *What Is Life? The Physical Aspect of the Living Cell* (Cambridge, UK: Cambridge University Press, 1944).

37 **Penrose explicitly rejects:** Roger Penrose, *The Emperor's New Mind: Concerning Computers, Minds, and the Laws of Physics* (New York: Oxford University Press, 1989).

37 **The biophysicist Francis Crick:** F. Crick and C. Koch, "Towards a neurobiological theory of consciousness," *Seminars in the Neurosciences* 2 (1990): 263–275; Francis Crick, *The Astonishing Hypothesis* (New York: Touchstone, 1994).

38 **It is the powerful cultural vestige:** Marleen Rozemond, *Descartes's Dualism* (Cambridge, MA: Harvard University Press, 1998).

38 **In Descartes's depiction:** René Descartes, *The Passions of the Soul*, translated by Stephen Voss (Indianapolis: Hackett, 1989).

38 **The ego and id:** Christopher Badcock, "Freud: Fraud or Folk-Psychologist?," *Psychology Today*, September 3, 2012; Saul McLeod, "Id, Ego and Superego," SimplyPsychology, www.simplypsychology.org/psyche.html, 2007.

39 **"integrate[s] cerebral":** Bandai, "Body and Brain Connection—Xbox 360," Amazon.com (accessed March 23, 2017).

40 **On a February morning in 1685:** Dorothy Senior, *The Gay King: Charles II, His Court and Times* (New York: Brentano's, 1911).

40 **An excess of blood:** An excess of blood was known as a "plethora" in the jargon of humorism.

41 **A fifth of the brain's volume:** Setti Rengachary and Richard Ellenbogen, eds., *Principles of Neurosurgery*, 2nd ed. (New York: Elsevier Mosby, 2004).

41 **The less noticed brain cells:** S. Herculano-Houzel, "The glia/neuron ratio: How it varies uniformly across brain structures and species and what that means for brain physiology and evolution," *Glia* 62 (2014): 1377–1391.

42 **This terrible disease:** Gina Kolata and Lawrence K. Altman, "Weighing Hope and Reality in Kennedy's Cancer Battle," *New York Times*, August 27, 2009.

42 **Many of these conditions:** Interestingly, the structure of the US National Institutes of Health almost codifies a vestigial mind-body separation within the government's infrastructure for neuromedicine and neuroscience research. Research into pathologies like stroke and concussion is overseen by a part of the NIH called the National Institute for Neurological Diseases and Stroke (NINDS). NINDS is distinct from the NIH institutes that deal with more cognitive brain disorders, the National Institute for Mental Health (NIMH) and the National Institute for Drug Addiction (NIDA).

42 **Calcium fluctuations:** N. Bazargani and D. Attwell, "Astrocyte calcium signaling: The third wave," *Nature Neuroscience* 19 (2016): 182–189.

42 **My MIT colleague Mriganka Sur:** J. Schummers, H. Yu, and M. Sur, "Tuned responses of astrocytes and their influence on hemodynamic signals in the visual cortex," *Science* 320 (2008): 1638–1643.

43 **Discovery of functional hyperemia:** Stefano Zago, Lorenzo Lorusso, Roberta Ferrucci, and Alberto Priori, "Functional Neuroimaging: A Historical Perspective," in *Neuroimaging: Methods*, edited by Peter Bright (Rijeka, Croatia: InTechOpen, 2012).

43 **Certain drugs that act:** G. Garthwaite et al., "Signaling from blood vessels to CNS axons through nitric oxide," *Journal of Neuroscience* 26 (2006): 7730–7740; E. Ruusuvuori and K. Kaila, "Carbonic anhydrases and brain pH in the control of neuronal excitability," *Subcellular Biochemistry* 75 (2014): 271–290.

43 **There are also hints:** C. I. Moore and R. Cao, "The hemo-neural hypothesis: On the role of blood flow in information processing," *Journal of Neurophysiology* 99 (2008): 2035–2047.

43 **Selective activation of glia:** M. Hausser, "Optogenetics: The age of light," *Nature Methods* 11 (2014): 1012–1014.

43 **In one example, Ko Matsui:** T. Sasaki et al., "Application of an optogenetic byway for perturbing neuronal activity via glial photostimulation," *Proceedings of the National Academy of Sciences* 109 (2012): 20720–20725.

44 **Her laboratory transplanted:** X. Han et al., "Forebrain engraftment by human glial progenitor cells enhances synaptic plasticity and learning in adult mice," *Cell Stem Cell* 12 (2013): 342–353.

45 **Perhaps most obviously, neurotransmitters:** Dale Purves, George J. Augustine, David Fitzpatrick, Lawrence C. Katz, Anthony-Samuel LaMantia, James

O. McNamara, and S. Mark Williams, eds., *Neuroscience*, 2nd ed. (Sunderland, MA: Sinauer Associates, 2001).

45 **In parts of the central nervous system:** John E. Dowling, *The Retina: An Approachable Part of the Brain* (Cambridge, MA: Belknap Press of Harvard University Press, 1987).

45 **The functional effects of gliotransmitters:** D. Li, C. Agulhon, E. Schmidt, M. Oheim, and N. Ropert, "New tools for investigating astrocyte-to-neuron communication," *Frontiers in Cellular Neuroscience* 7 (2013): 193.

46 **In doing so, the drugs:** J. O. Schenk, "The functioning neuronal transporter for dopamine: Kinetic mechanisms and effects of amphetamines, cocaine and methylphenidate," *Progress in Drug Research* 59 (2002): 111–131.

46 **Neurotransmitter diffusion also:** B. Barbour and M. Hausser, "Intersynaptic diffusion of neurotransmitter," *Trends in Neuroscience* 20 (1997): 377–384.

46 **A number of studies have documented:** N. Arnth-Jensen, D. Jabaudon, and M. Scanziani, "Cooperation between independent hippocampal synapses is controlled by glutamate uptake," *Nature Neuroscience* 5 (2002): 325–331; P. Marcaggi and D. Attwell, "Short- and long-term depression of rat cerebellar parallel fibre synaptic transmission mediated by synaptic crosstalk," *Journal of Physiology* 578 (2007): 545–550; Y. Okubo et al., "Imaging extrasynaptic glutamate dynamics in the brain," *Proceedings of the National Academy of Sciences* 107 (2010): 6526–6531.

46 **Instead, both synaptic cross-talk:** K. H. Taber and R. A. Hurley, "Volume transmission in the brain: Beyond the synapse," *Journal of Neuropsychiatry and Clinical Neuroscience* 26 (2014): iv, 1–4.

47 **Indeed, in the nervous systems:** S. R. Lockery and M. B. Goodman, "The quest for action potentials in *C. elegans* neurons hits a plateau," *Nature Neuroscience* 12 (2009): 377–378.

48 **"every aspect of thinking":** Douglas R. Hofstadter, *Gödel, Escher, Bach: An Eternal Golden Braid* (New York: Basic Books, 1979).

NOTES TO CHAPTER 3

51 **Few things in today's internet-dominated:** Users' comments in the online Urban Dictionary include: (1) "Refers to a couple in an ambiguous state between 'friends' and 'in a relationship.' May also be used to indicate dissatisfaction with an existing relationship." (2) "Any relationship that's not OK; fear of being called single; holding on to something that's about to end; still hoping to work things out; in denial stage of separation." (3) "[A] couple that can't decide to be friends, friends with benefits, or to be in a full out relationship." www.urbandictionary .com/define.php?term=It%27s+complicated (accessed March 25, 2017).

51 **"the most complex object":** Christof Koch, quoted in Ira Flatow, "Decoding 'the Most Complex Object in the Universe,'" *Talk of the Nation*, National Public Radio, June 14, 2013.

51 **"If our brains were":** David Eagleman, *Incognito: The Secret Lives of the Brain* (New York: Vintage Books, 2012).

52 **"No computer comes close":** Alun Anderson, "Brain Work," *Economist*, November 17, 2011.

52 **"We won't be able":** Robin Murray, quoted in Edi Stark, "The Brain Is the 'Most Complicated Thing in the Universe,'" *Stark Talk*, BBC Radio Scotland, May 28, 2012.

52 **"The human brain is":** Voltaire, quoted in Julian Cribb, "The Self-Deceiver (*Homo delusus*)," Chapter 9 in *Surviving the 21st Century: Humanity's Ten Great Challenges and How We Can Overcome Them* (Cham, Switzerland: Springer International, 2016).

52 **"The more complicated":** Brian Thomas, "Brain's Complexity 'Is Beyond Anything Imagined,'" Institute for Creation Research, discovercreation.org /blog/2013/12/20/brains-complexity-is-beyond-anything-imagined, January 17, 2011.

52 **The legendary Indian sage Vyasa:** *Krishna: The Beautiful Legend of God*, translated by Edwin F. Bryant (New York: Penguin, 2004).

53 **"fixed stars which almost":** Paul Lettinck, *Aristotle's Meteorology and Its Reception in the Arab World* (Boston: Brill, 1999).

53 **"By the aid of the telescope":** Galileo Galilei, *The Sidereal Messenger*, translated by Edward S. Carlos (London: Rivingtons, 1880).

53 **In his original sketches:** Reproduced in Stanley Finger, *Minds Behind the Brain: A History of the Pioneers and Their Discoveries* (New York: Oxford University Press, 2000).

54 **Similarly elaborate architectures:** Richard Rapport, *Nerve Endings: The Discovery of the Synapse* (New York: W. W. Norton, 2005).

54 **Catalog Aria:** W. A. Mozart and L. Da Ponte, *Don Giovanni* (New York: Ricordi, 1986).

54 **The ATLAS subatomic particle detector:** "ATLAS Fact Sheet," European Organization for Nuclear Research (CERN), 2011.

54 **The detector processes:** Fortunately for the purposes of data management, most events that occur in the ATLAS detector are rejected by the detector's triggering mechanisms, which reject all but about two hundred "interesting" events per second.

54 **We now know that Galileo's:** M. Temming, "How Many Stars Are There in the Universe?" *Sky & Telescope*, July 15, 2014.

54 **"billions and billions":** Carl Sagan, *Billions and Billions: Thoughts on Life and Death at the Brink of the Millennium* (New York: Ballantine Books, 1997).

54 **To take on the task:** S. Herculano-Houzel and R. Lent, "Isotropic fractionator: A simple, rapid method for the quantification of total cell and neuron numbers in the brain," *Journal of Neuroscience* 25 (2005): 2518–2521.

55 **Herculano-Houzel and her colleagues:** F. A. Azevedo et al., "Equal numbers of neuronal and nonneuronal cells make the human brain an isometrically scaled-up primate brain," *Journal of Comparative Neurology* 513 (2009): 532–541.

55 **By counting synapses this way:** J. DeFelipe, P. Marco, I. Busturia, and A. Merchan-Perez, "Estimation of the number of synapses in the cerebral cortex: Methodological considerations," *Cerebral Cortex* 9 (1999): 722–732.

55 **This type of process indicates:** Published estimates of the number of synapses per neuron vary considerably, with most sources reporting numbers in the range from one thousand to ten thousand, and some even exceeding this range.

55 **Genes are turned on:** Y. Ko et al., "Cell type-specific genes show striking and distinct patterns of spatial expression in the mouse brain," *Proceedings of the National Academy of Sciences* 110 (2013): 3095–3100.

56 **Mitochondria:** D. Attwell and S. B. Laughlin, "An energy budget for signaling in the grey matter of the brain," *Journal of Cerebral Blood Flow Metabolism* 21 (2001): 1133–1145.

56 **According to some estimates:** B. Pakkenberg et al., "Aging and the human neocortex," *Experimental Gerontology* 38 (2003): 95–99; "Table HM-20: Public Road Length, 2013, Miles by Functional System," Office of Highway Policy Information, Federal Highway Administration, www.fhwa.dot.gov /policyinformation/statistics/2013/hm20.cfm, October 21, 2014.

56 **For contrast, consider the liver:** E. Bianconi et al., "An estimation of the number of cells in the human body," *Annals in Human Biology* 40 (2013): 463–471.

57 **The task of mapping:** Sebastian Seung, *Connectome: How the Brain's Wiring Makes Us Who We Are* (Boston: Houghton Mifflin Harcourt, 2012).

57 **In one of the first published connectomics:** M. Helmstaedter et al., "Connectomic reconstruction of the inner plexiform layer in the mouse retina," *Nature* 500 (2013): 168–174; John E. Dowling, *The Retina: An Approachable Part of the Brain* (Cambridge, MA: Belknap Press of Harvard University Press, 1987).

58 **Normal adult brain sizes:** J. S. Allen, H. Damasio, and T. J. Grabowski, "Normal neuroanatomical variation in the human brain: an MRI-volumetric study," *American Journal of Physical Anthropology* 118 (2002): 341–358.

58 **Brain volume correlates:** A. W. Toga and P. M. Thompson, "Genetics of brain structure and intelligence," *Annual Review of Neuroscience* 28 (2005): 1–23.

58 **Although some of the disparities:** S. Herculano-Houzel, D. J. Messeder, K. Fonseca-Azevedo, and N. A. Pantoja, "When larger brains do not have more neurons: Increased numbers of cells are compensated by decreased average cell size across mouse individuals," *Frontiers in Neuroanatomy* 9 (2015): 64.

58 **Brain volume decreases:** N. C. Fox and J. M. Schott, "Imaging cerebral atrophy: Normal ageing to Alzheimer's disease," *Lancet* 363 (2004): 392–394.

58 **In 2014, a twenty-four-year-old:** F. Yu, Q. J. Jiang, X. Y. Sun, and R. W. Zhang, "A new case of complete primary cerebellar agenesis: Clinical and imaging findings in a living patient," *Brain* 138 (2015): e353.

59 **A group of surgeons:** E. P. Vining et al., "Why would you remove half a brain? The outcome of 58 children after hemispherectomy—the Johns Hopkins experience: 1968 to 1996," *Pediatrics* 100 (1997): 163–171.

60 **"I have witnessed":** C. C. Abbott, "Intelligence of the crow," *Science* 1 (1883): 576.

60 **More generally, members of the corvid:** N. J. Emery and N. S. Clayton, "The mentality of crows: Convergent evolution of intelligence in corvids and apes," *Science* 306 (2004): 1903–1907.

61 **"You turkey!":** Irene M. Pepperberg, *Alex & Me: How a Scientist and a Parrot Discovered a Hidden World of Animal Intelligence—and Formed a Deep Bond in the Process* (New York: HarperCollins, 2008).

61 **The punch line:** A. N. Iwaniuk, K. M. Dean, and J. E. Nelson, "Interspecific allometry of the brain and brain regions in parrots (psittaciformes): Comparisons with other birds and primates," *Brain, Behavior and Evolution* 65

(2005): 40–59; J. Mehlhorn, G. R. Hunt, R. D. Gray, G. Rehkamper, and O. Gunturkun, "Tool-making New Caledonian crows have large associative brain areas," *Brain, Behavior and Evolution* 75 (2010): 63–70.

61 **These animals could not:** S. Olkowicz et al., "Birds have primate-like numbers of neurons in the forebrain," *Proceedings of the National Academy of Sciences* 113 (2016): 7255–7260; S. Herculano-Houzel, "The remarkable, yet not extraordinary, human brain as a scaled-up primate brain and its associated cost," *Proceedings of the National Academy of Sciences* 109, Suppl 1 (2012): 10661–10668.

61 **Although we humans outrank:** G. Roth and U. Dicke, "Evolution of the brain and intelligence," *Trends in Cognitive Science* 9 (2005): 250–257.

62 **Yet the capybara's brain:** S. Herculano-Houzel, B. Mota, and R. Lent, "Cellular scaling rules for rodent brains," *Proceedings of the National Academy of Sciences* 103 (2006): 12138–12143; J. L. Kruger, N. Patzke, K. Fuxe, N. C. Bennett, and P. R. Manger, "Nuclear organization of cholinergic, putative catecholaminergic, serotonergic and orexinergic systems in the brain of the African pygmy mouse (*Mus minutoides*): Organizational complexity is preserved in small brains," *Journal of Chemical Neuroanatomy* 44 (2012): 45–56. The number of neurons in the African pygmy mouse is not published, but an estimate of fewer than 60 million neurons in the pygmy mouse brain can be obtained by applying Herculano-Houzel and colleagues' finding that brain size is proportional to neuron count raised to the power of 1.587 within the rodent family. Reference values of 71 million neurons and brain mass of 416 milligrams for mice were also obtained from Herculano-Houzel et al., and a brain mass of 275 milligrams was used for pygmy mice, as cited by Kruger et al.

62 **An ant's brain:** M. A. Seid, A. Castillo, and W. T. Wcislo, "The allometry of brain miniaturization in ants," *Brain, Behavior and Evolution* 77 (2011): 5–13.

63 **"the brain of an ant is":** Charles Darwin, *The Descent of Man, and Selection in Relation to Sex* (London: John Murray, 1871).

63 **Some biologists propose:** Harry J. Jerison, *Evolution of the Brain and Intelligence* (New York: Academic, 1973).

63 **In fact, many neuroscientists believe:** X. Jiang et al., "Principles of connectivity among morphologically defined cell types in adult neocortex," *Science* 350 (2015): aac9462.

63 **One such structure:** V. B. Mountcastle, "The columnar organization of the neocortex," *Brain* 120 (Part 4) (1997): 701–722.

64 **"What I cannot create,":** "Richard Feynman's Blackboard at Time of His Death," Caltech Image Archive, archives-dc.library.caltech.edu (accessed March 29, 2017).

64 **Some people have cited:** Sean Hill, "Whole Brain Simulation," in *The Future of the Brain*, edited by Gary Marcus and Jeremy Freeman (Princeton, NJ: Princeton University Press, 2015).

64 **The billion-dollar European initiative:** HBP-PS Consortium, *The Human Brain Project: A Report to the European Commission*, 2012.

64 **Related efforts in the United States:** A. P. Alivisatos et al., "The brain activity map project and the challenge of functional connectomics," *Neuron* 74 (2012): 970–974.

65 **One of the few organisms:** C. I. Bargmann and E. Marder, "From the connectome to brain function," *Nature Methods* 10 (2013): 483–490.

65 **Scientists today can measure:** Peter Shadbolt, "Scientists Upload a Worm's Mind into a Lego Robot," CNN, January 21, 2015.

66 **A car, after all:** Cars can admittedly also double as media players, climate conditioners, power sources, and sleeping quarters, but these functions are for the most part dispensable.

68 **A related attitude inspired:** Stephen J. Gould, *The Mismeasure of Man* (New York: W. W. Norton, 1996).

68 **Brain morphology was:** Ralph L. Holloway, Chet C. Sherwood, Patrick R. Hof, and James K. Rilling, "Evolution of the Brain in Humans—Paleoneurology," *Encyclopedia of Neuroscience*, edited by Marc D. Binder, Nobutaka Hirokawa, and Uwe Windhorst (Berlin, Germany: Springer, 2009).

68 **Neanderthals, who originated:** D. Falk et al., "The brain of LB1, *Homo floresiensis*," *Science* 308 (2005): 242–245.

68 **The earliest art:** J. DeFelipe, "The evolution of the brain, the human nature of cortical circuits, and intellectual creativity," *Frontiers in Neuroanatomy* 5 (2011): 29.

69 **The "uncontacted" tribes:** B. Holmes, "How many uncontacted tribes are there in the world?" *New Scientist*, August 22, 2013.

NOTES TO CHAPTER 4

71 **These procedures have:** The 1979 Nobel Prize in Physiology or Medicine was awarded to Allen Cormack and Godfrey Hounsfield "for the development of computer assisted tomography." The 2003 Nobel Prize in Physiology or Medicine was given to Paul Lauterbur and Peter Mansfield "for their discoveries concerning magnetic resonance imaging."

71 **Over ten thousand medical:** A PubMed search for "neuroimaging" returns an average of 10,039 articles per year over the five-year period spanning 2012–2016. A search for articles that include both "brain" and "imaging" returns an average of 17,270 articles per year for the same period.

72 **In the 1990s, fMRI emerged:** J. W. Belliveau et al., "Functional mapping of the human visual cortex by magnetic resonance imaging," *Science* 254 (1991): 716–719; S. Ogawa et al., "Intrinsic signal changes accompanying sensory stimulation: Functional brain mapping with magnetic resonance imaging," *Proceedings of the National Academy of Sciences* 89 (1992): 5951–5955.

72 **To perform an fMRI experiment:** S. A. Huettel, A. W. Song, and G. McCarthy, eds., *Functional Magnetic Resonance Imaging*, 3rd ed. (Sunderland, MA: Sinauer Associates, 2014).

72 **More edgy studies:** S. Schleim, T. M. Spranger, S. Erk, and H. Walter, "From moral to legal judgment: The influence of normative context in lawyers and other academics," *Social Cognitive and Affective Neuroscience* 6 (2011): 48–57; S. M. McClure et al., "Neural correlates of behavioral preference for culturally familiar drinks," *Neuron* 44 (2004): 379–387.

72 **"The influence of fMRI-based":** B. R. Rosen and R. L. Savoy, "fMRI at 20: Has it changed the world?," *NeuroImage* 62 (2012): 1316–1324.

72 **Hundreds of newspaper articles:** A LexisNexis Academic search for records containing the search term "fMRI" with content type and category set to "newspapers" retrieved 1,187 hits from the time period of April 1, 2013, to March 31, 2017.

72 **Readers get hooked:** Marco Iacobini, Joshua Freedman, and Jonas Kaplan, "This Is Your Brain on Politics," *New York Times*, November 11, 2007; Benedict Carey, "Watching New Love As It Sears the Brain," *New York Times*, May 31, 2005.

72 **At the same time:** Sally Satel and Scott O. Lilienfeld, *Brainwashed: The Seductive Appeal of Mindless Neuroscience* (New York: Basic Books, 2013).

73 **In a much discussed 2008 article:** D. P. McCabe and A. D. Castel, "Seeing is believing: The effect of brain images on judgments of scientific reasoning," *Cognition* 107 (2008): 343–352.

74 **Cognitive neuroscientists Cayce Hook and Martha Farah:** C. J. Hook and M. J. Farah, "Look again: Effects of brain images and mind-brain dualism on lay evaluations of research," *Journal of Cognitive Neuroscience* 25 (2013): 1397–1405.

75 **Over the past decade:** M. Kaufman, "Meditation Gives Brain a Charge, Study Finds," *Washington Post*, January 3, 2005.

75 **But Davidson and his colleagues found:** J. A. Brefczynski-Lewis, A. Lutz, H. S. Schaefer, D. B. Levinson, and R. J. Davidson, "Neural correlates of attentional expertise in long-term meditation practitioners," *Proceedings of the National Academy of Sciences* 104 (2007): 11483–11488.

75 **He says he is interested:** Sharon Begley, "How Thinking Can Change the Brain," *Wall Street Journal*, January 19, 2007.

75 **Davidson's work with the meditating:** D. Biello, "Searching for God in the brain," *Scientific American* 18 (2007): 38–45.

76 **In one study, Newberg's team:** A. B. Newberg, N. A. Wintering, D. Morgan, and M. R. Waldman, "The measurement of regional cerebral blood flow during glossolalia: A preliminary SPECT study," *Psychiatry Research* 148 (2006): 67–71.

76 **"I don't think faith":** V. Mabrey and R. Sherwood, "Speaking in Tongues: Alternative Voices in Faith," ABC News, March 20, 2007.

76 **A 2007 *Scientific American* article:** Biello, "Searching for God in the brain."

76 **"A wealth of scientific studies,":** Mario Beauregard, *Brain Wars: The Scientific Battle over the Existence of the Mind and the Proof That Will Change the Way We Live Our Lives* (New York: HarperCollins, 2012).

77 **Modern brain imaging was born:** *The Scanner Story*, directed by Michael Weigall, EMITEL Productions, 1977.

78 **Imaging these molecules:** M. M. Ter-Pogossian, M. E. Phelps, E. J. Hoffman, and N. A. Mullani, "A positron-emission transaxial tomograph for nuclear imaging (PETT)," *Radiology* 114 (1975): 89–98.

78 **In one approach:** A. Newberg, A. Alavi, and M. Reivich, "Determination of regional cerebral function with FDG-PET imaging in neuropsychiatric disorders," *Seminars in Nuclear Medicine* 32 (2002): 13–34.

78 **A second PET functional imaging:** Michael E. Phelps, *PET: Molecular Imaging and Its Biological Applications* (New York: Springer, 2004).

78 **A recent breakthrough:** W. E. Klunk et al., "Imaging brain amyloid in Alzheimer's disease with Pittsburgh Compound-B," *Annals of Neurology* 55 (2004): 306–319.

79 **Even the fastest functional PET:** Peter Doggers, "Magnus Carlsen Checkmates Bill Gates in 12 Seconds," Chess.com, chess.com/news/view/bill-gates -vs-magnus-carlsen-checkmate-in-12-seconds-8224, January 24, 2014.

79 **In the first published fMRI study:** Belliveau et al., "Functional mapping of the human visual cortex by magnetic resonance imaging."

79 **At about the same time:** S. Ogawa, T. M. Lee, A. R. Kay, and D. W. Tank, "Brain magnetic resonance imaging with contrast dependent on blood oxygenation," *Proceedings of the National Academy of Sciences* 87 (1990): 9868–9872; S. Ogawa et al., "Intrinsic signal changes accompanying sensory stimulation."

80 **Not surprisingly, limitations:** N. K. Logothetis, "What we can do and what we cannot do with fMRI," *Nature* 453 (2008): 869–878.

80 **Berkeley neuroimager Jack Gallant:** Elizabeth Landau, "Scan a Brain, Read a Mind?," CNN, April 12, 2014.

80 **This usually involves expansive:** William B. Penny, Karl J. Friston, John T. Ashburner, Stefan J. Kiebel, and Thomas E. Nichols, eds., *Statistical Parametric Mapping: The Analysis of Functional Brain Images* (New York: Academic, 2006).

81 **Instead, functional brain maps:** Modern bologna derives from the traditional northern Italian pork sausage known as mortadella. Although today's bologna products can be made with meats other than pork, they are all highly processed and distant from their corresponding animal sources, and even more so from pigs.

81 **A University of California:** C. M. Bennett, M. B. Miller, and G. L. Wolford, "Neural correlates of interspecies perspective taking in the post-mortem Atlantic Salmon: An argument for multiple comparisons correction," *Journal of Serendipitous and Unexpected Results* 1 (2010): 1–5.

81 **Bennett had a difficult time:** "About the Ig Nobel Prizes," Improbable Research, www.improbable.com/ig (accessed May 4, 2017).

81 **A second damning study:** E. Vul, C. Harris, P. Winkielman, and H. Pashler, "Puzzlingly high correlations in fMRI studies of emotion, personality, and social cognition," *Perspectives on Psychological Science* 4 (2009): 274–290.

82 **The brain is like a Swiss:** Nancy Kanwisher, "A Neural Portrait of the Human Mind," TED Conferences, March 19, 2014.

82 **Perhaps the best-known example:** C. W. Domanski, "Mysterious 'Monsieur Leborgne': The mystery of the famous patient in the history of neuropsychology is explained," *Journal of the History of Neuroscience* 22 (2013): 47–52.

84 **"What's important":** Kanwisher, "A Neural Portrait."

84 **"Critics feel that fMRI":** D. Dobbs, "Fact or phrenology?," *Scientific American* 16 (2005): 24.

84 **"One can be almost certain":** R. A. Poldrack, "Mapping mental function to brain structure: How can cognitive neuroimaging succeed?," *Perspectives on Psychological Science* 5 (2010): 753–761.

84 **Titles such as "Neural Correlates":** T. K. Inagaki and N. I. Eisenberger, "Neural correlates of giving support to a loved one," *Psychosomatic Medicine*

74 (2012): 3–7; C. Lamm, C. D. Batson, and J. Decety, "The neural substrate of human empathy: Effects of perspective-taking and cognitive appraisal," *Journal Cognitive Neuroscience* 19 (2007): 42–58; K. H. Lee et al., "Neural correlates of superior intelligence: Stronger recruitment of posterior parietal cortex," *NeuroImage* 29 (2006): 578–586.

85 **Advertising expert Martin Lindstrom:** Martin Lindstrom, "You Love Your iPhone. Literally," *New York Times*, September 30, 2011.

85 **He writes that "the aSTG":** Jonah Lehrer, *Imagine: How Creativity Works* (Boston: Houghton Mifflin, 2012).

85 **Even the Nobel Prize–winning biologist:** Francis Crick, *The Astonishing Hypothesis* (New York: Touchstone, 1994).

86 **A key reason:** Neuroskeptic, "Brain Scanning—Just the Tip of the Iceberg?," Neuroskeptic Blog, blogs.discovermagazine.com/neuroskeptic/2012/03/21/brain-scanning-just-the-tip-of-the-iceberg, March 21, 2012.

86 **By doing so, they:** J. V. Haxby et al., "Distributed and overlapping representations of faces and objects in ventral temporal cortex," *Science* 293 (2001): 2425–2430.

87 **It could even be the case:** A. Shmuel, M. Augath, A. Oeltermann, and N. K. Logothetis, "Negative functional MRI response correlates with decreases in neuronal activity in monkey visual area V1," *Nature Neuroscience* 9 (2006): 569–577.

87 **famous guard dog:** Arthur Conan Doyle, "The Adventure of Silver Blaze," in *The Memoirs of Sherlock Holmes* (London: George Newnes, 1894).

87 **"Even if we could associate":** William R. Uttal, *The New Phrenology: The Limits of Localizing Cognitive Processes in the Brain* (Cambridge, MA: MIT Press, 2003).

87 **In a similar vein, philosopher:** Daniel Dennett, *Consciousness Explained* (Boston: Back Bay Books, 1992).

88 **In Beckett's absurdist masterpiece:** Samuel Beckett, *Waiting for Godot: A Tragicomedy in Two Acts* (New York: Grove, 1954).

89 **"Claims that computational methods":** Logothetis, "What we can do and what we cannot do with fMRI."

89 **Some apparently specialized:** N. Kanwisher and G. Yovel, "The fusiform face area: A cortical region specialized for the perception of faces," *Philosophical Transactions of the Royal Society of London Series B: Biological Sciences* 361 (2006): 2109–2128.

89 **By combining cutting-edge:** M. B. Ahrens, M. B. Orger, D. N. Robson, J. M. Li, and P. J. Keller, "Whole-brain functional imaging at cellular resolution using light-sheet microscopy," *Nature Methods* 10 (2013): 413–420.

90 **Some of my own laboratory's:** B. B. Bartelle, A. Barandov, and A. Jasanoff, "Molecular fMRI," *Journal of Neuroscience* 36 (2016): 4139–4148.

NOTES TO CHAPTER 5

92 *Your Brain Is God:* Timothy Leary, *Your Brain Is God* (Berkeley, CA: Ronin, 2001).

92 **"All mental functions,":** Eric R. Kandel, "Your Mind Is Nothing but Neurons, and That's Fine," Big Think, www.bigthink.com/videos/a-biological-basis-for-the-unconscious (accessed May 5, 2017).

92 **" 'You' . . . are nothing":** Francis Crick, *The Astonishing Hypothesis* (New York: Touchstone, 1994).

92 **Reporting on a study:** Robert Lee Hotz, "A Neuron's Obsession Hints at Biology of Thoughts," *Wall Street Journal*, October 9, 2009.

92 **"Behind your thoughts":** Friedrich Nietzsche, *Thus Spake Zarathustra*, translated by Thomas Common (Buffalo, NY: Prometheus Books, 1993).

93 **Wittgenstein writes:** Ludwig Wittgenstein, *Philosophical Investigations*, translated by G. E. M. Anscombe (New York: Macmillan, 1953).

93 **"By speaking about the brain's":** Maxwell R. Bennett and Peter M. S. Hacker, *Philosophical Foundations of Neuroscience* (Malden, MA: Blackwell, 2003).

93 **"*I* feel pain,":** Daniel Dennett, "Philosophy as Naive Anthropology: Comment on Bennett and Hacker," in *Neuroscience and Philosophy: Brain, Mind, and Language*, edited by Maxwell Bennett et al. (New York: Columbia University Press, 2007).

93 **Other philosophers of mind:** Patricia Churchland, *Touching a Nerve: The Self as Brain* (New York: W. W. Norton, 2013); Derek Parfit, *Reasons and Persons* (New York: Oxford University Press, 1984).

94 **Approximately so ended:** R. S. Boyer, E. A. Rodin, T. C. Grey, and R. C. Connolly, "The skull and cervical spine radiographs of Tutankhamen: A critical appraisal," *American Journal of Neuroradiology* 24 (2003): 1142–1147.

95 **To the Egyptians:** A. A. Fanous and W. T. Couldwell, "Transnasal excerebration surgery in ancient Egypt," *Journal of Neurosurgery* 116 (2012): 743–748.

95 **A group called the Brain:** The Brain Preservation Foundation, www.brainpreservation.org (accessed May 5, 2017).

95 **Another organization, called Alcor:** Alcor Life Extension Foundation: The World's Leader in Cryonics, www.alcor.com (accessed May 5, 2017).

95 **"We are our brains,":** S. W. Bridge, "The neuropreservation option: Head first into the future," *Cryonics* 16 (1995): 4–7.

95 **Subjected to a battery of tests:** K. Hussein, E. Matin, and A. G. Nerlich, "Paleopathology of the juvenile Pharaoh Tutankhamun: 90th anniversary of discovery," *Virchows Archiv* 463 (2013): 475–479.

96 **"He might be envisioned":** Z. Hawass et al., "Ancestry and pathology in King Tutankhamun's family," *Journal of the American Medical Association* 303 (2010): 638–647.

96 **DNA evidence from the mummy:** World Health Organization Communicable Diseases Cluster, "Severe falciparum malaria," *Transactions of the Royal Society of Tropical Medicine and Hygiene* 94, Suppl 1 (2000): S1–90.

96 **As recently as:** Edward Shorter, *A History of Psychiatry: From the Era of the Asylum to the Age of Prozac* (New York: John Wiley & Sons, 1997).

97 **"Schumann was always ahead":** Hans-Joachim Kreuzer, Interview by Wolf-Dieter Seiffert, "Schumann's 'Late Works,'" Schumann Forum 2010, henleusa.com/en/schumann-anniversary-2010/schumann-forum/the-late-works.html.

97 **Psychiatrist Bradford Felker:** B. Felker, J. J. Yazel, and D. Short, "Mortality and medical comorbidity among psychiatric patients: A review," *Psychiatric Services* 47 (1996): 1356–1363.

99 **The biology behind:** Eric J. Nestler, Steven E. Hyman, David M. Holtzman, and Robert C. Malenka, *Molecular Neuropharmacology: A Foundation for Clinical Neuroscience* (New York: McGraw-Hill Education, 2015).

100 **The pupil dilation:** A. W. Tank and D. Lee Wong, "Peripheral and central effects of circulating catecholamines," *Comprehensive Physiology* 5 (2015): 1–15.

100 **Adrenaline and cortisol:** A. Schulz and C. Vogele, "Interoception and stress," *Frontiers in Psychology* 6 (2015): 993.

100 **These hormonal changes:** L. M. Glynn, E. P. Davis, and C. A. Sandman, "New insights into the role of perinatal HPA-axis dysregulation in postpartum depression," *Neuropeptides* 47 (2013): 363–370.

101 **An angry man, for instance:** Charles Darwin, *The Expression of the Emotions in Man and Animals* (London: John Murray, 1872).

101 **"We feel sorry":** William James, *The Principles of Psychology* (New York: Henry Holt and Company, 1890).

101 **In a survey of over a hundred:** Psychologist Lisa Feldman Barrett argues, on the other hand, that evidence for the specificity of physiological responses to emotions is overstated. She calls into question the definitions of emotional categories themselves, arguing that emotional responses are more fluid and variable than we commonly conceive. The kind of alternative perspective she proposes "would not deny the importance of evolutionarily preserved responses, but might deny emotions any privileged status as innate neural circuits or modules" (L. F. Barrett, *Perspectives on Psychological Science* 1 [2006]: 28–58). See also S. D. Kreibig, "Autonomic nervous system activity in emotion: A review," *Biological Psychiatry* 84 (2010): 394–421.

101 **A fascinating 2014 analysis:** L. Nummenmaa, E. Glerean, R. Hari, and J. K. Hietanen, "Bodily maps of emotions," *Proceedings of the National Academy of Sciences* 111 (2014): 646–651.

103 **"When a negative somatic":** Antonio Damasio, *Descartes' Error: Emotion, Reason, and the Human Brain* (New York: G. P. Putnam, 1994).

103 **The motivation for Damasio's:** A. R. Damasio, "The somatic marker hypothesis and the possible functions of the prefrontal cortex," *Philosophical Transactions of the Royal Society of London Series B: Biological Sciences* 351 (1996): 1413–1420.

103 **Although some critics question:** B. D. Dunn, T. Dalgleish, and A. D. Lawrence, "The somatic marker hypothesis: A critical evaluation," *Neuroscience & Biobehavioral Reviews* 30 (2006): 239–271.

104 **"The opportunities for bodily":** Joseph E. LeDoux, *The Emotional Brain: The Mysterious Underpinnings of Emotional Life* (New York: Simon & Schuster, 1996).

104 **"operates automatically":** Daniel Kahneman, *Thinking, Fast and Slow* (New York: Farrar, Straus and Giroux, 2011).

104 **The legendary violinist:** In a 1978 article in the *Journal of the American Medical Association* (27: 141–162), Myron Schonenfeld speculated that Paganini most likely experienced Marfan Syndrome, a rare genetic disorder.

104 **"His hand was not larger":** F. Bennati, quoted in A. Pedrazzini, A. Martelli, and S. Tocco, "Niccolò Paganini: The hands of a genius," *Acta Biomedica* 86 (2015): 27–31.

104 **He designed fabulously virtuosic:** Carl Guhr, *Paganini's Art of Playing the Violin: With a Treatise on Single and Double Harmonic Notes*, translated by S. Novello (London: Novello & Co., 1915).

105 **Even Carl Gauss:** W. K. Bühler, *Gauss: A Biographical Study* (New York: Springer, 1981).

105 **"mathematics is the product":** George Lakoff and Rafael E. Núñez, *Where Mathematics Comes From: How the Embodied Mind Brings Mathematics into Being* (New York: Basic Books, 2000).

105 **"Our bodies and":** A. D. Wilson and S. Golonka, "Embodied cognition is not what you think it is," *Frontiers in Psychology* 4 (2013): 58.

106 **Each beaver has:** L. M. Gordon et al., "Dental materials: Amorphous intergranular phases control the properties of rodent tooth enamel," *Science* 347 (2015): 746–750.

106 **It is an instinct:** E. N. Woodcock, *Fifty Years a Hunter and a Trapper* (St. Louis: A. R. Harding, 1913).

106 **In her book *Beyond the Brain*:** Louise Barrett, *Beyond the Brain: How Body and Environment Shape Animal and Human Minds* (Princeton, NJ: Princeton University Press, 2011).

107 **This is because our environments:** James J. Gibson, "The Theory of Affordances," in *Perceiving, Acting, and Knowing: Toward an Ecological Psychology*, edited by Robert Shaw and John Bransford (Hillsdale, NJ: Lawrence Erlbaum Associates, 1977).

107 **"Because so many of the concepts":** George Lakoff and Mark Johnson, *Metaphors We Live By* (Chicago: University of Chicago Press, 1980).

108 **In one example:** A. Eerland, T. M. Guadalupe, and R. A. Zwaan, "Leaning to the left makes the Eiffel Tower seem smaller: Posture-modulated estimation," *Psychological Science* 22 (2011): 1511–1514.

108 **In another demonstration:** L. K. Miles, L. K. Nind, and C. N. Macrae, "Moving through time," *Psychological Science* 21 (2010): 222–223.

108 **In adults over fifty:** J. M. Northey, N. Cherbuin, K. L. Pumpa, D. J. Smee, and B. Rattray, "Exercise interventions for cognitive function in adults older than 50: A systematic review with meta-analysis," *British Journal of Sports Medicine* (2017).

108 **There is some evidence:** E. P. Cox et al., "Relationship between physical activity and cognitive function in apparently healthy young to middle-aged adults: A systematic review," *Journal of Science and Medicine in Sport* 19 (2016): 616–628.

109 **In one of the most impressive:** M. Oppezzo and D. L. Schwartz, "Give your ideas some legs: The positive effect of walking on creative thinking," *Journal of Experimental Psychology: Learning, Memory, and Cognition* 40 (2014): 1142–1152.

109 **Exercise is now known:** K. Weigmann, "Why exercise is good for your brain: A closer look at the underlying mechanisms suggests that some sports, especially combined with mental activity, may be more effective than others," *EMBO Reports* 15 (2014): 745–748.

110 **"I woke up knowing":** Claire Sylvia with William Novak, *A Change of Heart: A Memoir* (New York: Warner Books, 1997).

110 **A recent article:** Joe Shute, "The Life-Saving Operations That Change Personalities," *Telegraph*, February 6, 2015.

110 **"Heart transplants trigger":** Will Oremus, "Personality Transplant," *Slate*, March 26, 2012.

110 **A team of Austrian researchers:** B. Bunzel, B. Schmidl-Mohl, A. Grundbock, and G. Wollenek, "Does changing the heart mean changing personality? A retrospective inquiry on 47 heart transplant patients," *Quality of Life Research* 1 (1992): 251–256.

111 **Replacing the diseased organ:** M. E. Olbrisch, S. M. Benedict, K. Ashe, and J. L. Levenson, "Psychological assessment and care of organ transplant patients," *Journal of Consulting and Clinical Psychology* 70 (2002): 771–783.

111 **A rare before/after comparison:** K. Mattarozzi, L. Cretella, M. Guarino, and A. Stracciari, "Minimal hepatic encephalopathy: Follow-up 10 years after successful liver transplantation," *Transplantation* 93 (2012): 639–643.

111 **Major components of the network:** Michael D. Gershon, *The Second Brain: The Scientific Basis of Gut Instinct and a Groundbreaking New Understanding of Nervous Disorders of the Stomach and Intestine* (New York: HarperCollins, 1998).

111 **It turns out that:** S. Fass, "Gastric Sleeve Surgery—The Expert's Guide," Obesity Coverage, obesitycoverage.com, April 13, 2017.

112 **"I am bombarded":** H. Woodberries, "Personality Changes—It's a Huge Deal!!" Gastric Sleeve Discussion Forum, gastricsleeve.com (March 10, 2012).

112 **The prevalence of such:** Jeff Seidel, "After Bariatric Surgery, the Rules of Marriage Often Change," *Seattle Times*, June 1, 2011.

112 **Bacteriotherapy methods:** CDC Newsroom, "Nearly Half a Million Americans Suffered from *Clostridium difficile* Infections in a Single Year," Centers for Disease Control and Prevention (February 25, 2015).

112 **And recent research supports:** Peter A. Smith, "Can the Bacteria in Your Gut Explain Your Mood?" *New York Times*, June 23, 2015; T. G. Dinan, R. M. Stilling, C. Stanton, and J. F. Cryan, "Collective unconscious: How gut microbes shape human behavior," *Journal of Psychiatric Research* 63 (2015): 1–9.

112 **In one experiment:** P. Bercik et al., "The intestinal microbiota affect central levels of brain-derived neurotropic factor and behavior in mice," *Gastroenterology* 141 (2011): 599–609.

113 **In a different experiment:** J. A. Bravo et al., "Ingestion of *Lactobacillus* strain regulates emotional behavior and central GABA receptor expression in a mouse via the vagus nerve," *Proceedings of the National Academy of Sciences* 108 (2011): 16050–16055.

113 **Both the Collins:** K. Tillisch et al., "Consumption of fermented milk product with probiotic modulates brain activity," *Gastroenterology* 144 (2013): 1394–1401.

113 **Gage was "no longer Gage,":** J. M. Harlow, "Recovery from the passage of an iron bar through the head," *Publication of the Massachusetts Medical Society* 2 (1869): 327–347; A. Bechara, H. Damasio, D. Tranel, and A. R. Damasio, "The Iowa Gambling Task and the somatic marker hypothesis: Some questions and answers," *Trends in Cognitive Science* 9 (2005): 159–162; discussion 62–64.

114 **Over thirty thousand people worldwide:** J. Horgan, "The forgotten era of brain chips," *Scientific American* 293 (2005): 66–73.

114 **The development of optogenetics:** J. Gorman, "Brain Control in a Flash of Light," *New York Times*, April 21, 2014.

114 **Blocking adenosine:** W. R. Lovallo et al., "Caffeine stimulation of cortisol secretion across the waking hours in relation to caffeine intake levels," *Psychosomatic Medicine* 67 (2005): 734–739; J. R. Schwartz and T. Roth, "Neurophysiology of sleep and wakefulness: Basic science and clinical implications," *Current Neuropharmacology* 6 (2008): 367–378.

NOTES TO CHAPTER 6

118 **"Increasingly it is recognized":** Peter M. Milner, *The Autonomous Brain: A Neural Theory of Attention and Learning* (Mahwah, NJ: Lawrence Erlbaum, 1999).

118 **University College London:** P. Haggard, "Human volition: Towards a neuroscience of will," *Nature Reviews Neuroscience* 9 (2008): 934–946.

118 **Part of the blame:** B. Libet, C. A. Gleason, E. W. Wright, and D. K. Pearl, "Time of conscious intention to act in relation to onset of cerebral activity (readiness-potential): The unconscious initiation of a freely voluntary act," *Brain* 106 (Part 3) (1983): 623–642.

118 **As neuroscientist David Eagleman:** David Eagleman, *Incognito: The Secret Lives of the Brain* (New York: Vintage Books, 2012).

119 **The 2015 Disney hit:** *Inside Out*, directed by Pete Docter and Ronnie Del Carmen (Walt Disney Studios, 2015).

119 **The resulting contradiction:** Gilbert Ryle, *The Concept of Mind* (New York: Hutchinson's University Library, 1949).

120 **In his famous essay:** Arthur Schopenhauer, *Prize Essay on the Freedom of the Will*, translated by E. F. J. Payne (New York: Cambridge University Press, 1999).

120 **The oldest example:** *Fodor's Tokyo*, edited by Stephanie E. Butler (New York: Random House, 2011).

120 **Figurines of the monkeys:** Three-Monkeys, three-monkeys.info (accessed May 10, 2017).

121 **A statuette of the monkeys:** Juhi Saklani, *Eyewitness Gandhi* (New York: DK Publishing, 2014).

121 **Elements of the Italian mafia:** Neil Strauss, "Mafia Songs Break a Code of Silence; A Gory Italian Folk Form Attracts Fans, and Critics," *New York Times*, July 22, 2002.

121 **It is telling:** *The Bhagavad Gita*, translated by Laurie L. Patton (New York: Penguin Classics, 2008).

121 **By measuring electrical signals:** H. B. Barlow, W. R. Levick, and M. Yoon, "Responses to single quanta of light in retinal ganglion cells of the cat," *Vision Research*, Suppl 3 (1971): 87–101.

122 **Caltech professor Markus Meister:** M. Meister, R. O. Wong, D. A. Baylor, and C. J. Shatz, "Synchronous bursts of action potentials in ganglion cells of the developing mammalian retina," *Science* 252 (1991): 939–943.

122 **In one study, researchers:** K. Koch et al., "How much the eye tells the brain," *Current Biology* 16 (2006): 1428–1434.

122 **Most of the auditory neurons:** B. C. Moore, "Coding of sounds in the auditory system and its relevance to signal processing and coding in cochlear implants," *Otology & Neurotology* 24 (2003): 243–254.

122 **Some touch receptors:** R. S. Johansson and A. B. Vallbo, "Tactile sensibility in the human hand: Relative and absolute densities of four types of mechanoreceptive units in glabrous skin," *Journal of Physiology* 286 (1979): 283–300.

122 **The olfactory receptor:** Daniel L. Schacter, Daniel T. Gilbert, Daniel M. Wegner, and Matthew K. Nock, *Psychology*, 3rd ed. (New York: Worth Publishers, 2014).

122 **This means that despite:** T. Connelly, A. Savigner, and M. Ma, "Spontaneous and sensory-evoked activity in mouse olfactory sensory neurons with defined odorant receptors," *Journal of Neurophysiology* 110 (2013): 55–62.

123 **This much data directed:** Eric Griffith, "How Fast Is Your Internet Connection . . . Really?" *PC Magazine*, June 2, 2017.

123 **Most of the brain's motor:** E. V. Evarts, "Relation of Discharge Frequency to Conduction Velocity in Pyramidal Tract Neurons," *Journal of Neurophysiology* 28 (1965): 216–228; L. Firmin et al., "Axon diameters and conduction velocities in the macaque pyramidal tract," *Journal of Neurophysiology* 112 (2014): 1229–1240.

124 **More than 40 percent:** David C. Van Essen, "Organization of Visual Areas in Macaque and Human Cerebral Cortex," in *Visual Neurosciences*, vol. 1, edited by Leo M. Chalupa and John S. Werner (Cambridge, MA: MIT Press, 2004).

124 **neural signals in the visual:** N. Naue et al., "Auditory event–related response in visual cortex modulates subsequent visual responses in humans," *Journal of Neuroscience* 31 (2011): 7729–7736.

124 **For example, researchers have:** C. Kayser, C. I. Petkov, and N. K. Logothetis, "Multisensory interactions in primate auditory cortex: fMRI and electrophysiology," *Hearing Research* 258 (2009): 80–88.

124 **Brain areas known for:** Micah M. Murray and Mark T. Wallace, eds., *The Neural Bases of Multisensory Processes* (Boca Raton, FL: CRC, 2012).

124 **Visual responses in frontal regions:** M. T. Schmolesky et al., "Signal timing across the macaque visual system," *Journal of Neurophysiology* 79 (1998): 3272–3278.

124 **A remarkable phenomenon:** M. E. Raichle et al., "A default mode of brain function," *Proceedings of the National Academy of Sciences* 98 (2001): 676–682.

125 **To study brain dynamics:** B. Biswal, F. Z. Yetkin, V. M. Haughton, and J. S. Hyde, "Functional connectivity in the motor cortex of resting human brain using echo-planar MRI," *Magnetic Resonance in Medicine* 34 (1995): 537–541.

125 **Such correlations are thought:** K. R. Van Dijk et al., "Intrinsic functional connectivity as a tool for human connectomics: Theory, properties, and optimization," *Journal of Neurophysiology* 103 (2010): 297–321.

125 **Neurologist Maurizio Corbetta:** V. Betti et al., "Natural scenes viewing alters the dynamics of functional connectivity in the human brain," *Neuron* 79 (2013): 782–797.

125 **Tamara Vanderwal of Yale:** T. Vanderwal, C. Kelly, J. Eilbott, L. C. Mayes, and F. X. Castellanos, "Inscapes: A movie paradigm to improve compliance in functional magnetic resonance imaging," *NeuroImage* 122 (2015): 222–232.

126 **Another study, led by:** N. Gaab, J. D. Gabrieli, and G. H. Glover, "Resting in peace or noise: Scanner background noise suppresses default-mode network," *Human Brain Mapping* 29 (2008): 858–867.

126 **Even the most banal:** J. H. Kaas, "The evolution of neocortex in primates," *Progress in Brain Research* 195 (2012): 91–102.

126 **"All I could feel":** Albert Camus, *The Stranger*, translated by Matthew Ward (New York: Vintage, 1989).

127 **"Meursault's crime seems":** Matthew H. Bowker, "Meursault and Moral Freedom: *The Stranger*'s Unique Challenge to an Enlightenment Ideal," in *Albert Camus's The Stranger: Critical Essays*, edited by Peter Francev (Newcastle upon Tyne, UK: Cambridge Scholars, 2014).

127 **In a 1994 study:** A. Vrij, J. van der Steen, and L. Koppelaar, "Aggression of police officers as a function of temperature: An experiment with the fire arms training system," *Journal of Community & Applied Social Psychology* 4 (1994): 365–370.

127 **In an ambitious survey of sixty:** S. M. Hsiang, M. Burke, and E. Miguel, "Quantifying the influence of climate on human conflict," *Science* 341 (2013): 123567.

127 **In one instance, the number:** E. G. Cohn and J. Rotton, "Assault as a function of time and temperature: A moderator-variable time-series analysis," *Journal of Personality and Social Psychology* 72 (1997): 1322–1334.

127 **Backing this up:** L. Taylor, S. L. Watkins, H. Marshall, B. J. Dascombe, and J. Foster, "The impact of different environmental conditions on cognitive function: A focused review," *Frontiers in Physiology* 6 (2015): 372.

128 **Even untrained laboratory mice:** G. Greenberg, "The effects of ambient temperature and population density on aggression in two inbred strains of mice, *Mus musculus*," *Behaviour* 42 (1972): 119–130.

128 **Appreciation of this phenomenon:** Caroline Overy and E. M. Tansey, eds., *The Recent History of Seasonal Affective Disorder (SAD): The Transcript of a Witness Seminar*, Wellcome Witnesses to Contemporary Medicine, vol. 51 (London: Queen Mary, University of London, 2014).

128 **This finding was repeated:** N. E. Rosenthal et al., "Seasonal affective disorder: A description of the syndrome and preliminary findings with light therapy," *Archives of General Psychiatry* 41 (1984): 72–80; A. Magnusson, "An overview of epidemiological studies on seasonal affective disorder," *Acta Psychiatrica Scandinavica* 101 (2000): 176–184; K. A. Roecklein and K. J. Rohan, "Seasonal affective disorder: An overview and update," *Psychiatry* 2 (2005): 20–26.

129 **Ambient light levels control:** G. Pail et al., "Bright-light therapy in the treatment of mood disorders," *Neuropsychobiology* 64 (2011): 152–162.

129 **There are competing theories:** Roecklein and Rohan, "Seasonal affective disorder."

129 **"Colour is a means":** Wassily Kandinsky, *On the Spiritual in Art* (New York: Solomon R. Guggenheim Foundation, 1946).

129 **Biological studies of the effects:** Michael York, *The A to Z of New Age Movements* (Lanham, MD: Scarecrow, 2009).

129 **Pleasonton's method:** A. J. Pleasonton, *The Influence of the Blue Ray of the Sunlight and of the Blue Colour of the Sky; in Developing Animal and Vegetable Life, in Arresting Disease and in Restoring Health in Acute and Chronic Disorders to Human and Domestic Animals* (Philadelphia: Claxton, Remsen & Haffelfinger, 1876).

129 **A notable example:** Adam Alter, *Drunk Tank Pink: And Other Unexpected Forces That Shape How We Think, Feel, and Behave* (New York: Penguin, 2014).

130 **He convinced a local prison:** A. G. Schauss, "Tranquilizing effect of color reduces aggressive behavior and potential violence," *Orthomolecular Psychiatry* 8 (1979): 218–221.

130 **The fact that further experiments:** J. E. Gilliam and D. Unruh, "The effects of Baker-Miller pink on biological, physical and cognitive behaviour," *Journal of Orthomolecular Medicine* 3 (1988): 202–206.

130 **In a rigorous study of the effects:** P. Valdez and A. Mehrabian, "Effects of color on emotions," *Journal of Experimental Psychology: General* 123 (1994): 394–409.

130 **In one example, researchers at:** A. J. Elliot, M. A. Maier, A. C. Moller, R. Friedman, and J. Meinhardt, "Color and psychological functioning: The effect of red on performance attainment," *Journal of Experimental Psychology: General* 136 (2007): 154–168.

130 **A 2009 paper published in *Science*:** R. Mehta and R. J. Zhu, "Blue or red? Exploring the effect of color on cognitive task performances," *Science* 323 (2009): 1226–1229.

131 **Psychology researchers have defined:** P. Salamé and A. D. Baddeley, "Disruption of short-term memory by unattended speech: Implications for the structure of working memory," *Journal of Verbal Learning & Verbal Behavior* 21 (1982): 150–164; D. M. Jones and W. J. Macken, "Irrelevant tones produce an irrelevant speech effect: Implications for phonological coding in working memory," *Journal of Experimental Psychology* 19 (1993): 369–381.

132 **In one example of the effect:** E. M. Elliott, "The irrelevant-speech effect and children: Theoretical implications of developmental change," *Memory and Cognition* 30 (2002): 478–487.

132 **Researchers at the University of London:** S. Murphy and P. Dalton, "Out of touch? Visual load induces inattentional numbness," *Journal of Experimental Psychology: Human Perception and Performance* 42 (2016): 761–765.

132 **In another study, scientists:** S. Brodoehl, C. M. Klingner, and O. W. Witte, "Eye closure enhances dark night perceptions," *Science Reports* 5 (2015): 10515.

132 **One of the most bizarre:** H. McGurk and J. MacDonald, "Hearing lips and seeing voices," *Nature* 264 (1976): 746–748. The McGurk effect was first demonstrated by presenting video and auditory stimuli to a group of subjects, and then surveying them about their perceptions. You can now experience the effect yourself using video files available online.

133 **Neuroscientists convey this distinction:** M. Corbetta and G. L. Shulman, "Control of goal-directed and stimulus-driven attention in the brain," *Nature Reviews Neuroscience* 3 (2002): 201–215.

133 **The great William James wrote:** William James, *The Principles of Psychology* (New York: Henry Holt and Company, 1890).

133 **In bottom-up attention:** R. J. Krauzlis, A. Bollimunta, F. Arcizet, and L. Wang, "Attention as an effect not a cause," *Trends in Cognitive Science* 18 (2014): 457–464.

134 **Many neuroscientists believe:** Corbetta and Shulman, "Control of goal-directed and stimulus-driven attention in the brain."

134 **How this happens:** M. Handford, *Where's Waldo? The Complete Collection* (Cambridge, MA: Candlewick, 2008).

134 **Research by attention expert:** N. P. Bichot, A. F. Rossi, and R. Desimone, "Parallel and serial neural mechanisms for visual search in macaque area V4," *Science* 308 (2005): 529–534.

134 **Our eye motions are:** H. F. Credidio, E. N. Teixeira, S. D. Reis, A. A. Moreira, and J. S. Andrade Jr., "Statistical patterns of visual search for hidden objects," *Scientific Reports* 2 (2012): 920.

134 **Our gaze jumps around:** I. Mertens, H. Siegmund, and O. J. Grusser, "Gaze motor asymmetries in the perception of faces during a memory task," *Neuropsychologia* 31 (1993): 989–998.

135 **The taste of a cookie:** Marcel Proust, À la recherche du temps perdu: Du côté de chez Swann, 7 vols. (Paris, France: Gallimard, 1919–1927); Evelyn Waugh, *Brideshead Revisited: The Sacred and Profane Memories of Captain Charles Ryder* (Boston: Little, Brown, 1945).

135 **The ancient historian Suetonius:** Gaius Suetonius Tranquillus, *The Twelve Caesars*, translated by Robert Graves (New York: Penguin, 2007).

135 **Neuroscientist John Medina:** John Medina, *Brain Rules: 12 Principles for Surviving and Thriving at Work, Home, and School* (Seattle: Pear, 2008).

135 **A widely reported study:** Alyson Gausby, *Attention Spans*, Consumer Insights, Microsoft Canada, 2015.

136 **In 1951, a young psychologist:** Solomon E. Asch, "Effects of Group Pressure upon the Modification and Distortion of Judgments," in *Groups, Leadership and Men: Research in Human Relations*, edited by H. Guetzkow (Oxford, UK: Carnegie, 1951).

136 **"That we have found":** S. E. Asch, "Opinions and social pressure," *Scientific American* (November 1955).

137 **Functional brain imaging studies have:** H. C. Breiter et al., "Response and habituation of the human amygdala during visual processing of facial expression," *Neuron* 17 (1996): 875–887.

137 **A famous example is contagious:** A. J. Bartholomew and E. T. Cirulli, "Individual variation in contagious yawning susceptibility is highly stable and largely unexplained by empathy or other known factors," *PLoS One* 9 (2014): e91773.

137 **Another example is the phenomenon:** S. Kouider and E. Dupoux, "Subliminal speech priming," *Psychological Science* 16 (2005): 617–625.

137 **For instance, the word "cow":** S. Kouider, V. de Gardelle, S. Dehaene, E. Dupoux, and C. Pallier, "Cerebral bases of subliminal speech priming," *NeuroImage* 49 (2010): 922–929.

137 **Today, tens of thousands:** Atul Gawande, "Hellhole," *New Yorker*, March 30, 2009.

137 **Prisoners in solitary confinement:** Sal Rodriguez, "Solitary Confinement: FAQ," Solitary Watch, solitarywatch.com/facts/faq, March 31, 2012.

138 **According to the advocacy group:** Sal Rodriguez, "Fact Sheet: Psychological Effects of Solitary Confinement," Solitary Watch, solitarywatch.com/facts /fact-sheets, June 4, 2011.

138 **"All day, every day":** Shruti Ravindran, "Twilight in the Box," *Aeon*, February 27, 2014.

138 **A 1972 study:** P. Gendreau, N. L. Freedman, G. J. Wilde, and G. D. Scott, "Changes in EEG alpha frequency and evoked response latency during solitary confinement," *Journal of Abnormal Psychology* 79 (1972): 545–549.

138 **You may have read:** Jan Harold Brunvand, ed., *American Folklore: An Encyclopedia* (New York: Garland, 1996).

138 **For instance, French people:** *Extramarital Affairs Topline*, Pew Research Center, 2014.

138 **It is extremely unlikely:** R. Khan, "Genetic map of Europe; genes vary as a function of distance," *Gene Expression*, May 21, 2008.

139 **Neuroscientist Michael Gazzaniga:** Michael Gazzaniga, *Who's in Charge? Free Will and the Science of the Brain* (New York: HarperCollins, 2012).

139 **It is perhaps in a similar:** John Donne, *Devotions upon Emergent Occasions and Death's Duel* (New York: Vintage, 1999).

NOTES TO CHAPTER 7

144 **While President Barack Obama:** Barack Obama, Remarks at the NAACP Conference, Philadelphia, 2015.

144 **"reject the idea":** Ronald Reagan, Speech at the Republican National Convention, Platform Committee Meeting, Miami, 1968.

145 **The cyclical nature:** Marx's view of history as a succession of class struggles is most famously set forth in *The Communist Manifesto*, coauthored with Friedrich Engels and published anonymously as a German edition in 1848.

146 **In the political realm:** Karl Marx, "Economic and Philosophic Manuscripts of 1844," in *Economic and Philosophic Manuscripts of 1844; and the Communist Manifesto*, translated by Martin Milligan (Buffalo, NY: Prometheus Books, 1988).

146 **In the late nineteenth century:** John M. O'Donnell, *The Origins of Behaviorism: American Psychology, 1870–1920* (New York: New York University Press, 1985).

147 **In his youth, Wundt:** S. Diamond, "Wundt Before Leipzig," in *Wilhelm Wundt and the Making of a Scientific Psychology*, edited by R. W. Rieber (New York: Springer, 1980).

147 **"In psychology,":** W. Wundt, "Principles of Physiological Psychology," translated by S. Diamond, in *Wilhelm Wundt and the Making of a Scientific Psychology*, edited by R. W. Rieber (New York: Plenum, 1980).

147 **Tools of the trade:** K. Danziger, "The history of introspection revisited," *Journal of the History of Behavioral Sciences* 16 (1980): 241–262.

148 **For instance, Wundt noticed:** W. Wundt, *Lectures on Human and Animal Psychology*. translated by J. E. Creighton and E. B. Titchener (New York: Macmillan, 1896).

148 **According to Wundt:** Wundt, *Principles of Physiological Psychology*.

148 **Consideration of brain physiology:** W. M. Wundt, *Outlines of Psychology*, translated by C. H. Judd (Leipzig, Germany: W. Engelman, 1897).

148 **"Experimental introspection is":** Edward B. Titchener, *A Primer of Psychology* (New York: Macmillan, 1899).

149 **He was the scion:** William James's father, Henry James Sr., was a noted theologian, and his siblings included the novelist Henry James and the diarist Alice James.

149 **The hairsplitting experimental:** William James, *The Principles of Psychology* (New York: Henry Holt and Company, 1890).

149 **"Introspective Observation,":** James, *The Principles of Psychology*.

149 **Their ideas spread:** A. Kim, "Wilhelm Maximilian Wundt," *The Stanford Encyclopedia of Philosophy*, edited by Edward N. Zalta (Stanford, CA: Metaphysics Research Lab, Center for the Study of Language and Information, Stanford University).

149 **The teachings of psychology's:** Prominent examples included Oswald Külpe at the University of Würtzburg, G. Stanley Hall at Johns Hopkins and then Clark University, Edward Thorndike and James Cattell at Columbia, Edwin Boring at Harvard, and Charles Spearman at University College London.

150 **Yerkes argued that:** R. M. Yerkes, "Eugenic bearing of measurements of intelligence in the United States Army," *Eugenics Review* 14 (1923): 225–245.

150 **This activism led Cattell:** C. S. Gruber, "Academic freedom at Columbia University, 1917–1918: The case of James McKeen Cattell," *AAUP Bulletin* 58 (1972): 297–305.

151 **"Psychology as the behaviorist views it":** J. B. Watson, "Psychology as the behaviorist views it," *Psychological Review* 20 (1913): 158–177.

152 **Even earlier research:** D. N. Robinson, *An Intellectual History of Psychology* (Madison: University of Wisconsin Press, 1986).

152 **Titchener regarded:** E. B. Titchener, "On 'Psychology as the behaviorist views it,'" *Proceedings of the American Philosophical Society* 53 (1914): 1–17.

152 **Robert Yerkes sharply:** R. M. Yerkes, "Comparative psychology: A question of definitions," *The Journal of Philosophy, Psychology and Scientific Methods* 10, 1913: 580–582; J. B. Watson, *Behavior: An Introduction to Comparative Psychology* (New York: Henry Holt and Company, 1914).

152 **Historian Franz Samelson:** F. Samelson, "Struggle for scientific authority: The reception of Watson's behaviorism, 1913–1920," *Journal of History of the Behavioral Sciences* 17 (1981): 399–425.

152 **Meanwhile, the patient-focused:** This chapter paints the history of psychology in admittedly broad strokes and cannot give due attention to many who have played significant parts in the field. Freud in particular could merit more space than I have allotted. One might regard his emphasis on the structure of the unconscious individual mind as generally consonant with the outlook of the introspective psychologists of the late nineteenth century, albeit considerably less granular and "scientific" in orientation. Academic psychologists largely rejected Freud, but a meeting point nevertheless occurred through the person of G. Stanley Hall, a student of both James and Wundt who engaged intellectually with psychological theories across the spectrum. Hall hosted

Freud's only visit to America in 1909. Hall himself also delved into such topics as paranormal psychology and the mental analysis of religious figures.

152 **If Watson was the Moses:** Michael Specter, "Drool," *New Yorker*, November 24, 2014.

153 **Pavlov most notably studied:** Daniel P. Todes, *Ivan Pavlov: A Russian Life in Science* (New York: Oxford University Press, 2014).

153 **In this way:** We saw examples of classical conditioning in Chapter 2, in which Schultz and colleagues paired a visual stimulus with a juice reward as part of their studies on dopamine neuron function during learning in monkeys, and where Nedergaard and colleagues paired a tone with an aversive shock to test learning ability in mice that had received transplanted human glial cells.

153 **In a self-acknowledged:** John B. Watson, *Behaviorism* (New York: W. W. Norton, 1925).

153 **The leading spokesperson:** S. J. Haggbloom et al., "The 100 most eminent psychologists of the 20th century," *Review of General Psychology* 6 (2002): 139–152.

153 **Skinner promoted:** Skinner elaborated on the phenomenon of operant conditioning most famously in his book *The Behavior of Organisms: An Experimental Analysis* (New York: Appleton-Century-Crofts, 1938). The concept is often attributed, however, to Edward Thorndike, who explained action-based learning as arising from what he termed the Law of Effect: "Of several responses made to the same situation, those which are accompanied or closely followed by satisfaction to the animal will, other things being equal, be more firmly connected with the situation, so that, when it recurs, they will be more likely to recur; those which are accompanied or closely followed by discomfort to the animal will, other things being equal, have their connections with that situation weakened, so that, when it recurs, they will be less likely to occur. The greater the satisfaction or discomfort, the greater the strengthening or weakening of the bond" (Edward L. Thorndike, *Animal Intelligence* [New York: Macmillan, 1911]).

154 **Skinner viewed operant:** B. F. Skinner, *Science and Human Behavior* (New York: Macmillan, 1953). A host of variants of operant conditioning were also introduced and studied by Skinner's contemporaries. Examples include Edwin Guthrie's contiguous conditioning, Edward Tolman's latent learning, and Murray Sidman's operant avoidance.

154 **A student of Skinner's:** Benedict Carey, "Sidney W. Bijou, Child Psychologist, Is Dead at 100," *New York Times*, July 21, 2009.

154 **Bijou's methods helped:** D. M. Baer, M. M. Wolf, and T. R. Risley, "Some current dimensions of applied behavior analysis," *Journal of Applied Behavioral Analysis* 1 (1968): 91–97.

154 **Offshoots of ABA:** J. J. Pear, "Behaviorism in North America since Skinner: A personal perspective," *Operants* Q4 (2015): 10–14.

154 **Students would answer questions:** J. Ludy and T. Benjamin, "A history of teaching machines," *American Psychologist* 43 (1988): 703–712.

155 **Writing in 1923:** Le Corbusier, *Toward an Architecture*, translated by John Goodman (Los Angeles: Getty Research Institute, 2007).

155 **A number of communal:** B. F. Skinner, *Walden Two* (New York: Macmillan, 1948).

155　**These settlements followed:** A. Sanguinetti, "The design of intentional communities: A recycled perspective on sustainable neighborhoods," *Behavior and Social Issues* 21 (2012): 5–25.

155　**As Watson put it:** Watson, *Behaviorism.*

155　**"We don't need to learn":** B. F. Skinner, quoted in Temple Grandin, *Animals in Translation: Using the Mysteries of Autism to Decode Animal Behavior* (New York: Scribner, 2005).

156　**"It was great for you,":** John Searle, quoted in Steven R. Postrel and Edward Feser, "Reality Principles: An Interview with John R. Searle," *Reason* (February 2000).

156　**Skinner had contended:** B. F. Skinner, *Verbal Behavior* (New York: Appleton-Century-Crofts, 1957).

157　**Chomsky scornfully dismissed:** Noam Chomsky, "A review of B. F. Skinner's *Verbal Behavior*," *Language* 35 (1959): 26–58.

157　**Conversely, Chomsky concluded:** Chomsky, "A review of B. F. Skinner's *Verbal Behavior.*"

158　**Psychologist Steven Pinker:** Steven Pinker, *The Blank Slate: The Modern Denial of Human Nature* (New York: Viking, 2002).

158　**Chomsky himself championed:** Noam Chomsky, *Aspects of the Theory of Syntax* (Cambridge, MA: MIT Press, 1965).

158　**"The theory of human nature":** Pinker, *The Blank Slate.*

159　**It was at this interface:** M. Rescorla, "The computational theory of mind," *The Stanford Encyclopedia of Philosophy*, edited by Edward N. Zalta (Stanford, CA: Metaphysics Research Lab, Center for the Study of Language and Information, Stanford University, n.d.).

159　**Marr famously characterized:** David Marr, *Vision: A Computational Investigation into the Human Representation and Processing of Visual Information* (San Francisco: W. H. Freeman, 1982).

159　**Captivated by such:** Ned Block, "The Mind as the Software of the Brain," in *Thinking*, vol. 3 of *An Invitation to Cognitive Science*, 2nd ed., edited by Daniel N. Osherson et al. (Cambridge, MA: MIT Press, 1995).

159　**Neuroscience became a hot:** Pinker, *The Blank Slate.*

159　**With behaviorism dethroned:** Notable holdouts against completely brain-centered views of mental function include those associated with the embodied cognition movement we considered in Chapter 5, as well as others who have emphasized the importance of brain-body interactions involving such phenomena as Damasio's somatic markers or body-wide stress responses.

159　**Some commentators:** Peter B. Reiner, "The Rise of Neuroessentialism," in *Oxford Handbook of Neuroethics*, edited by Judy Illes and Barbara J. Sahakian (New York: Oxford University Press, 2011).

159　**"Many of us overtly":** A. Roskies, "Neuroethics for the new millennium," *Neuron* 35 (2002): 21–23.

160　**The world's most notorious icon:** Alex Hannaford, "The Mysterious Vanishing Brains," *Atlantic*, December 2, 2014.

160　**In another part of the brain:** C. D. Chenar, "Charles Whitman Autopsy Report," Cook Funeral Home, Austin, TX, 1966.

161　**"In many ways Whitman":** Gary M. Lavergne, *A Sniper in the Tower:*

The Charles Whitman Murders (Denton: University of North Texas Press, 1997).

161 **When postmortem examination:** Governor's Committee and Invited Consultants, *Report to the Governor, Medical Aspects, Charles J. Whitman Catastrophe*, Austin, TX, 1966.

161 **Many of Whitman's friends:** Lavergne, *A Sniper in the Tower*.

161 **Amygdala expert Joseph LeDoux:** Joseph LeDoux, "Inside the Brain, Behind the Music, Part 5," The Beautiful Brain Blog, thebeautifulbrain.com/2010/07/ledoux-amydaloids-crime-of-passion, July 23, 2010.

161 **Eagleman predicts:** David Eagleman, "The Brain on Trial," *Atlantic* (July/August 2011).

162 **"Lawyers routinely order scans":** Jeffrey Rosen, "The Brain on the Stand," *New York Times Magazine*, March 11, 2007.

162 **Stanford neurobiologist Robert Sapolsky:** R. M. Sapolsky, "The frontal cortex and the criminal justice system," *Philosophical Transactions of the Royal Society of London Series B: Biological Sciences* 359 (2004): 1787–1796.

162 **Roughly thirty years later:** Adam Voorhes and Alex Hannaford, *Malformed: Forgotten Brains of the Texas State Mental Hospital* (New York: Powerhouse Books, 2014).

162 **"It's a mystery worthy":** Hannaford, "The Mysterious Vanishing Brains."

162 **Amid widespread media:** Rick Jervis and Doug Stanglin, "Mystery of Missing University of Texas Brains Solved," *USA Today*, December 3, 2014.

164 **When we strive for explanations:** Mary Midgley, *The Myths We Live By* (New York: Routledge, 2003).

164 **"Youth is hot":** William Shakespeare, "The Passionate Pilgrim," in *The Complete Works*, edited by Stephen Orgel and A. R. Braunmuller (New York: Penguin Books, 2002).

165 **"Regions within the limbic system":** Sarah-Jayne Blakemore, "The Mysterious Workings of the Adolescent Brain," TED Conferences, September 17, 2012.

165 **We might also note:** Maggie Koerth-Baker, "Who Lives Longest?" *New York Times Magazine*, March 19, 2013.

166 **Stressing the centrality:** "The Science of Drug Abuse and Addiction: The Basics," National Institute of Drug Abuse, drugabuse.gov/publications/media-guide/science-drug-abuse-addiction-basics (accessed June 7, 2017).

166 **Part of NIDA's objective:** A. I. Leshner, "Addiction is a brain disease, and it matters," *Science* 278 (1997): 45–47.

166 **External social and environmental variables:** J. D. Hawkins, R. F. Catalano, and J. Y. Miller, "Risk and protective factors for alcohol and other drug problems in adolescence and early adulthood: Implications for substance abuse prevention," *Psychological Bulletin* 112 (1992): 64–105; Mayo Clinic Staff, "Drug Addiction: Risk Factors," Mayo Clinic, mayoclinic.org/diseases-conditions/drug-addiction/basics/risk-factors/con-20020970 (accessed June 7, 2017).

166 **Sally Satel and Scott Lilienfeld argue:** Sally Satel and Scott O. Lilienfeld, *Brainwashed: The Seductive Appeal of Mindless Neuroscience* (New York: Basic Books, 2013).

166 **He writes that "addictive acts":** Lance Dodes, "Is Addiction Really a Disease?" *Psychology Today*, December 17, 2011.

166 **The hero of Mel Brooks's:** *Young Frankenstein*, directed by Mel Brooks, 20th Century Fox, 1974.

167 **Science writer Brian Burrell:** Brian Burrell, *Postcards from the Brain Museum: The Improbable Search for Meaning in the Matter of Famous Minds* (New York: Broadway Books, 2004).

167 **Modern researchers apply:** Nancy C. Andreasen, "Secrets of the Creative Brain," *Atlantic* (July/August 2014).

167 **According to studies of creativity:** M. Reznikoff, G. Domino, C. Bridges, and M. Honeyman, "Creative abilities in identical and fraternal twins," *Behavioral Genetics* 3 (1973): 365–377; A. A. Vinkhuyzen, S. van der Sluis, D. Posthuma, and D. I. Boomsma, "The heritability of aptitude and exceptional talent across different domains in adolescents and young adults," *Behavioral Genetics* 39 (2009): 380–392; C. Kandler et al., "The nature of creativity: The roles of genetic factors, personality traits, cognitive abilities, and environmental sources," *Journal of Personality & Social Psychology* 111 (2016): 230–249.

167 **Psychologist Kevin Dunbar:** Kevin Dunbar, "How Scientists Think: Online Creativity and Conceptual Change in Science," in *Creative Thought: An Investigation of Conceptual Structures and Processes*, edited by Thomas B. Ward, Steven M. Smith, and Jyotsna Vaid (Washington, DC: American Psychological Association, 1997).

167 **"A change in perspective":** Maria Konnikova, *Mastermind: How to Think Like Sherlock Holmes* (New York: Viking, 2013).

168 **In 1819 Franz Gall:** Jan Verplaetse, *Localising the Moral Sense: Neuroscience and the Search for the Cerebral Seat of Morality, 1800–1930* (New York: Springer, 2009).

168 **Reflecting a more recent view:** L. Pascual, P. Rodrigues, and D. Gallardo-Pujol, "How does morality work in the brain? A functional and structural perspective of moral behavior," *Frontiers in Integrative Neuroscience* 7 (2013): 65.

168 **This was stunningly:** S. Milgram, "Behavioral study of obedience," *Journal of Abnormal and Social Psychology* 67 (1963): 371–378.

169 **Psychologist Joshua Greene:** Lauren Cassani Davis, "Do Emotions and Morality mix?" *Atlantic*, February 5, 2016.

169 **The Scottish historian Thomas Carlyle:** Thomas Carlyle, *On Heroes, Hero-Worship, and the Heroic in History* (London: James Fraser, 1841).

169 **It is a picture that:** William James, "Great Men, Great Thoughts, and the Environment," *Atlantic Monthly* (October 1880).

NOTES TO CHAPTER 8

171 **"Schizophrenia is a disease":** K. Weir, "The roots of mental illness," *Monitor on Psychology* 43 (2012): 30.

171 **There is evidence:** G. Schomerus et al., "Evolution of public attitudes about mental illness: A systematic review and meta-analysis," *Acta Psychiatrica Scandinavica* 125 (2012): 440–452.

172 **According to the French social theorist:** Michel Foucault, *Madness and Civilization: A History of Insanity in the Age of Reason*, translated by Richard Howard (New York: Vintage Books, 1988).

172 **"To encourage the influence":** Samuel Tuke, *Description of the Retreat, an Institution near York, for Insane Persons of the Society of Friends: Containing an Account of Its Origins and Progress, the Modes of Treatment, and a Statement of Cases* (York, UK: Isaac Peirce, 1813).

172 **According to Foucault's analysis:** Foucault, *Madness and Civilization.*

172 **According to statistics:** "Mental Health Facts in America," National Alliance on Mental Illness, 2015, nami.org/Learn-More/Mental-Health-By-the -Numbers.

172 **On July 23, 2012:** Doug Stanglin, "Aurora Suspect James Holmes Sent His Doctor Burned Money," *USA Today,* December 10, 2012.

174 **And instead of using:** James Holmes, Laboratory Notebook, University of Colorado, 2012.

174 **After his arrest:** Ann O'Neill, Ana Cabrera, and Sara Weisfeldt, "A Look Inside the 'Broken' Mind of James Holmes," CNN, June 10, 2017.

175 **Since his teen years:** Ann O'Neill and Sara Weisfeldt, "Psychiatrist: Holmes Thought 3–4 Times a Day About Killing," CNN, June 10, 2017.

175 **For a more balanced perspective:** Jack Bragen, *Schizophrenia: My 35-Year Battle* (Raleigh, NC: Lulu, 2015).

175 **"In order to take medications,":** Jack Bragen, "On Mental Illness: The Sacrifices of Being Medicated," *Berkeley Daily Planet,* May 11, 2011.

175 **Self-stigmatizing is:** P. W. Corrigan, J. E. Larson, and N. Rusch, "Self-stigma and the 'why try' effect: Impact on life goals and evidence-based practices," *World Psychiatry* 8 (2009): 75–81.

175 **A large international analysis:** Schomerus et al., "Evolution of public attitudes about mental illness: A systematic review and meta-analysis."

176 **He and his collaborator:** P. W. Corrigan and A. C. Watson, "At issue: Stop the stigma: Call mental illness a brain disease," *Schizophrenia Bulletin* 30 (2004): 477–479.

176 **The twentieth century saw:** P. R. Reilly, "Eugenics and involuntary sterilization: 1907–2015," *Annual Review of Genomics and Human Genetics* 16 (2015): 351–368.

176 **A notorious example:** Dana Goldstein, "Sterilization's Cruel Inheritance," *New Republic,* March 4, 2016.

176 **In upholding the decision:** Carrie Buck v. John Hendren Bell, 274 U.S. 200 (1927).

177 **An unusual burial ceremony:** J. Pfeiffer, "Neuropathology in the Third Reich," *Brain Pathology* 1 (1991): 125–131.

177 **These specimens embodied:** Henry Friedlander, *The Origins of Nazi Genocide: From Euthanasia to the Final Solution* (Chapel Hill: University of North Carolina Press, 1997).

177 **Among those who benefitted:** J. T. Hughes, "Neuropathology in Germany during World War II: Julius Hallervorden (1882–1965) and the Nazi programme of 'euthanasia,'" *Journal of Medical Biography* 15 (2007): 116–122.

177 **Hallervorden was reputed:** J. Pfeiffer, "Phases in the postwar German Reception of the 'euthanasia program' (1939–1945) involving the killing of the mentally disabled and its exploitation by neuroscientists," *Journal of the History of the Neurosciences* 15 (2006): 210–244.

177 **That Hallervorden's brain samples:** R. Ahren, "German Institute Finds Brain Parts Used by Nazis for Research During, and After, WWII," *Times of Israel,* August 31, 2016.

178 **This reflex:** Roy Porter, "Madness and Its Institutions," in *Medicine in Society: Historical Essays,* edited by Andrew Wear (Cambridge, UK: Cambridge University Press, 1992).

178 **Stately neoclassical or gothic:** Mark Davis, *Asylum: Inside the Pauper Lunatic Asylums* (Stroud, UK: Amberley, 2014).

178 **But many treatment practices:** H. R. Rollin, "Psychiatry in Britain one hundred years ago," *British Journal of Psychiatry* 183 (2003): 292–298.

179 **A haunting 1869 photograph:** Chris Pleasance, "Faces from the Asylum: Harrowing Portraits of Patients at Victorian 'Lunatic' Hospital Where They Were Treated for 'Mania, Melancholia and General Paralysis of the Insane,'" *Daily Mail,* March 18, 2015.

179 **To modern eyes:** Ezra Susser, Sharon Schwartz, Alfredo Morabia, and Evelyn J. Bromet, eds., *Psychiatric Epidemiology: Searching for the Causes of Mental Disorders* (New York: Oxford University Press, 2006).

179 **Whereas most mental patients:** W. S. Bainbridge, "Religious insanity in America: The official nineteenth-century theory," *Sociological Analysis* 45 (1984).

180 **The most devastating condition:** G. Davis, "The most deadly disease of asylumdom: General paralysis of the insane and Scottish psychiatry, c. 1840–1940," *Journal of the Royal College of Physicians of Edinburgh* 42 (2012): 266–273.

180 **In an 1826 report:** J. M. S. Pearce, "Brain disease leading to mental illness: A concept initiated by the discovery of general paralysis of the insane," *European Neurology* 67 (2012): 272–278.

180 **By one account:** J. Hurn, "The changing fortunes of the general paralytic," *Wellcome History* 4 (1997): 5.

180 **On the Continent:** *Pellagra and Its Prevention and Control in Major Emergencies,* World Health Organization, 2000.

180 **One of the worst outbreaks:** Charles S. Bryan, *Asylum Doctor: James Woods Babcock and the Red Plague of Pellagra* (Columbia: University of South Carolina Press, 2014).

180 **A Hungarian-American epidemiologist:** V. P. Sydenstricker, "The history of pellagra, its recognition as a disorder of nutrition and its conquest," *American Journal of Clinical Nutrition* 6 (1958): 409–414.

181 **This is a brand of complexity:** Ludwik Fleck, *Genesis and Development of a Scientific Fact,* translated by Fred Bradley and Thaddeus J. Trenn (Chicago: University of Chicago Press, 1979).

181 **Correlating mental illnesses:** Stephen V. Faraone, Stephen J. Glatt, and Ming T. Tsuang, "Genetic Epidemiology," in *Textbook of Psychiatric Epidemiology,* edited by Ming T. Tsuang, Mauricio Tohen, and Peter B. Jones (Hoboken, NJ: John Wiley & Sons, 2011).

181 **If a person with a mental:** R. Plomin, M. J. Owen, and P. McGuffin, "The genetic basis of complex human behaviors," *Science* 264 (1994): 1733–1739.

182 **For instance, in studies:** Judith Allardyce and Jim van Os, "Examining

Gene-Environment Interplay in Psychiatric Disorders," in Tsuang, Tohen, and Jones, eds., *Textbook of Psychiatric Epidemiology*.

182 **Using approaches like these:** M. Burmeister, M. G. McInnis, and S. Zollner, "Psychiatric genetics: Progress amid controversy," *Nature Reviews Genetics* 9 (2008): 527–540.

182 **Based on reputable data:** P. F. Sullivan, M. J. Daly, and M. O'Donovan, "Genetic architectures of psychiatric disorders: The emerging picture and its implications," *Nature Reviews Genetics* 13 (2012): 537–551.

183 **Genes that correlate:** Supporting this point, a large 2014 study (F. A. Wright et al., "Heritability and genomics of gene expression in peripheral blood," *Nature Genetics* 46 [2014]: 430–437) found that about 70 percent of genes implicated in heritable autism spectrum disorders or mental retardation were associated with gene expression changes that could be detected in blood, a possible extracerebral source of physiological effects on mental function. A 2013 review of studies on the relationship between obesity and depression in adolescents (D. Nemiary et al., "The relationship between obesity and depression among adolescents," *Psychiatric Annual* 42 [2013]: 305–308), for instance, reports that obese adolescents are more likely than others to experience school and mental health problems, and that teasing and body dissatisfaction are both likely to be significant contributing factors. Because obesity in turn is linked to genetic causes, the obesity-depression relationship provides an illustration of how genes may influence the brain and mind through indirect means.

183 **For instance, traumatic brain:** M. Schwarzbold et al., "Psychiatric disorders and traumatic brain injury," *Neuropsychiatric Disease and Treatment* 4 (2008): 797–816.

183 **One such scholar:** Ruth Shonle Cavan, *Suicide* (Chicago: University of Chicago Press, 1928).

183 **A young graduate student:** J. Faris, "Robert E. Lee Faris and the discipline of sociology," *ASA Footnotes* 26 (1998): 8.

184 **These data revealed:** Robert E. L. Faris and H. Warren Dunham, *Mental Disorders in Urban Areas: An Ecological Study of Schizophrenia and Other Psychoses* (Chicago: University of Chicago Press, 1939).

185 **Toward the middle of the century:** A. V. Horwitz and G. N. Grob, "The checkered history of American psychiatric epidemiology," *Milbank Quarterly* 89 (2011): 628–657.

185 **Faris and Dunham were criticized:** J. D. Page, "Review of *Mental Disorders in Urban Areas*," *Journal of Educational Psychology* 30 (1939): 706–708.

185 **But subsequent studies:** W. W. Eaton, "Residence, social class, and schizophrenia," *Journal of Health and Social Behavior* 15 (1974): 289–299.

185 **Although some hypothesized:** Monica Charalambides, Craig Morgan, and Robin M. Murray, "Epidemiology of Migration and Serious Mental Illness: The Example of Migrants to Europe," in Tsuang, Tohen, and Jones, eds., *Textbook of Psychiatric Epidemiology*; G. Lewis, A. David, S. Andreasson, and P. Allebeck, "Schizophrenia and city life," *Lancet* 340 (1992): 137–140; M. Marcelis, F. Navarro-Mateu, R. Murray, J. P. Selten, and J. van Os, "Urbanization and

psychosis: A study of 1942–1978 birth cohorts in the Netherlands," *Psychological Medicine* 28 (1998): 871–879.

185 **Epidemiologists have found:** William W. Eaton, Chuan-Yu Chen, and Evelyn J. Bromet, "Epidemiology of Schizophrenia," in Tsuang, Tohen, and Jones, eds., *Textbook of Psychiatric Epidemiology*.

185 **Major depression, meanwhile:** D. S. Hasin, M. C. Fenton, and M. M. Weissman, "Epidemiology of Depressive Disorders," in Tsuang, Tohen, and Jones, eds., *Textbook of Psychiatric Epidemiology*.

185 **Bipolar disorder, which shares:** Kathleen R. Merikangas and Mauricio Tohen, "Epidemiology of Bipolar Disorder in Adults and Children," in Tsuang, Tohen, and Jones, eds., *Textbook of Psychiatric Epidemiology*.

186 **The writer Elie Wiesel:** Elie Wiesel, *A Mad Desire to Dance*, translated by Catherine Temerson (New York: Alfred A. Knopf, 2009).

186 **In Sylvia Plath's:** Sylvia Plath, *The Bell Jar* (London: Faber, 1966).

186 **Dostoevsky's Raskolnikov:** Fyodor Dostoyevsky, *Crime and Punishment*, translated by Oliver Ready (New York: Penguin, 2014).

186 **Most famously, Shakespeare's:** Wandering crazed on the stormy heath in Act III, Scene 2, Lear seems to invite explicit mental torment from the elements: "That have with two pernicious daughters join'd / Your high engender'd battles 'gainst a head / So old and white as this."

187 **On July 7, 1970:** "The Trial of Natalya Gorbanevskaya," *A Chronicle of Current Events*, August 31, 1970.

187 **Now she was accused:** The Editors, "Voices from the Past: The Trial of Gleb Pavlovsky," translated by J. Crowfoot, *A Chronicle of Events*, December 31, 1982.

187 **She was at that time:** "The Arrest of Natalya Gorbanevskaya," *A Chronicle of Current Events*, December 31, 1969.

187 **Experts at the Serbsky:** "The Trial of Natalya Gorbanevskaya."

187 *Sluggish schizophrenia*: H. Merskey and B. Shafran, "Political hazards in the diagnosis of 'sluggish schizophrenia,'" *British Journal of Psychiatry* 148 (1986): 247–256.

187 **"If my daughter":** "The Trial of Natalya Gorbanevskaya."

188 **Gorbanevskaya served two years:** Sidney Bloch and Peter Reddaway, *Russia's Political Hospitals: The Abuse of Psychiatry in the Soviet Union* (London: Futura, 1978).

188 **She was released in 1972:** Douglas Martin, "Natalya Gorbanevskaya, Soviet Dissident and Poet, Dies at 77," *New York Times*, December 1, 2013.

188 **The popular folksinger:** R. Apps, G. Moore, and S. Guppy, "Natalia," in *Joan Baez: From Every Stage* (A & M Records, 1976).

188 **The practice of politically motivated:** R. van Voren, "Political abuse of psychiatry—An historical overview," *Schizophrenia Bulletin* 36 (2010): 33–35.

188 **A Russian émigré:** Anne Applebaum, *Gulag: A History* (New York: Doubleday, 2003).

188 **In a *New York Times* piece:** Walter Reich, "The World of Soviet Psychiatry," *New York Times*, January 30, 1983.

188 **Where once such standards:** F. Jabr, "The newest edition of psychiatry's 'Bible,' the DSM-5, is complete," *Scientific American*, January 28, 2013.

189 **In its fifth and latest edition:** *The People Behind DSM-5*, American Psychiatric Association, 2013.

189 **The criteria continue:** *Diagnostic and Statistical Manual of Mental Disorders: DSM-5*, 5th ed. (Arlington, VA: American Psychiatric Association, 2013).

189 **The shifting of disease categories:** A. Suris, R. Holliday, and C. S. North, "The evolution of the classification of psychiatric disorders," *Behavioral Science* 6 (2016): 5.

189 **By looking around the globe:** Ethan Watters, *Crazy Like Us* (New York: Free Press, 2010).

190 **Watters notes that cross-cultural:** Ethan Watters, "The Americanization of Mental Illness," *New York Times Magazine*, January 8, 2010.

191 **Szasz explained:** T. Szasz, "The myth of mental illness," *American Psychology* 15 (1960): 113–118.

191 **Szasz earned notoriety:** J. Oliver, "The myth of Thomas Szasz," *New Atlantis* (Summer 2006); Benedict Carey, "Dr. Thomas Szasz, Psychiatrist Who Led Movement Against His Field, Dies at 92," *New York Times*, September 11, 2012.

191 **These could include so-called:** A. L. Petraglia, E. A. Winkler, and J. E. Bailes, "Stuck at the bench: Potential natural neuroprotective compounds for concussion," *Surgical Neurology International* 2 (2011): 146.

191 **Alternatively, if we chose:** Claude Quétel, *History of Syphilis*, translated by Judith Braddock and Brian Pike (Baltimore: Johns Hopkins University Press, 1990).

192 **Two poles of the debate:** G. L. Engel, "The need for a new medical model: A challenge for biomedicine," *Science* 196 (1977): 129–136.

192 **Engel was not neutral:** T. M. Brown, "George Engel and Rochester's Biopsychosocial Tradition: Historical and Developmental Perspectives," in *The Biopsychosocial Approach: Past, Present, and Future*, edited by Richard M. Frankel, Timothy E. Quill, and Susan H. McDaniel (Rochester, NY: University of Rochester Press, 2003).

192 **Doing this, he wrote:** Engel, "The need for a new medical model."

192 **The contrast between:** "Mental Health Treatment & Services," National Alliance on Mental Illness, nami.org/Learn-More/Treatment (accessed June 13, 2017).

192 **Examples include psychoanalytic:** Oliver Burkeman, "Therapy Wars: The Revenge of Freud," *Guardian*, January 7, 2016.

192 **Many drugs target:** J. R. Cooper, F. E. Bloom, and R. H. Roth, *The Biochemical Basis of Neuropharmacology* (New York: Oxford University Press, 2003).

192 **Other psychotherapeutic drugs:** G. S. Malhi and T. Outhred, "Therapeutic mechanisms of lithium in bipolar disorder: Recent advances and current understanding," *CNS Drugs* 30 (2016): 931–949.

193 **A widely cited report:** *America's State of Mind*, Medco Health Solutions, 2011.

193 **A similar study in England:** S. Ilyas and J. Moncrieff, "Trends in prescriptions and costs of drugs for mental disorders in England, 1998–2010," *British Journal of Psychiatry* 200 (2012): 393–398.

193 **Meanwhile, a study:** M. Olfson and S. C. Marcus, "National trends in outpatient psychotherapy," *American Journal of Psychiatry* 167 (2010): 1456–1463.

193 **The same surveys:** Schomerus et al., "Evolution of public attitudes about mental illness: A systematic review and meta-analysis."

193 **They write that "the brain-disease":** S. Satel and S. O. Lilienfeld, "Addiction and the brain-disease fallacy," *Frontiers in Psychiatry* 4 (2013): 141.

193 **He says that patients:** Robert Whitaker, *Anatomy of an Epidemic: Magic Bullets, Psychiatric Drugs, and the Astonishing Rise of Mental Illness in America* (New York: Crown Publishers, 2010).

194 **"The biological/psychosocial treatment":** A. Prosser, B. Helfer, and S. Leucht, "Biological v. psychosocial treatments: A myth about pharmacotherapy v. psychotherapy," *British Journal of Psychiatry* 208 (2016): 309–311.

195 **"Instead of arguing":** Corrigan and Watson, "At issue: Stop the stigma."

195 **"Technology can cover":** Antonio Regalado, "Why America's Top Mental Health Researcher Joined Alphabet," *Technology Review*, September 21, 2015.

195 **He suggests in particular:** The Google division Insel joined was subsequently spun off as the health sciences company Verily. Insel left Verily in 2017 and cofounded a company called Mindstrong, which also aims to use information technology, with the aid of smartphones, to diagnose and monitor mental illnesses.

196 **Psychologists Adrian Ward:** A. F. Ward and P. Valdesolo, "What internet habits say about mental health," *Scientific American*, August 14, 2012.

NOTES TO CHAPTER 9

197 **Before there was Kal-El:** Whitney Ellsworth, Robert J. Maxwell, and Bernard Luber, *Adventures of Superman*, Warner Bros. Television, September 19, 1952; Jerome Siegel and Joe Shuster, *The Reign of the Superman*, January 1933.

198 **Two years later:** Deborah Friedell, "Kryptonomics," *New Yorker*, June 24, 2013.

198 **Among the wonders:** P. Frati et al., "Smart drugs and synthetic androgens for cognitive and physical enhancement: Revolving doors of cosmetic neurology," *Current Neuropharmacology* 13 (2015): 5–11; K. Smith, "Brain decoding: Reading minds," *Nature* 502 (2013): 428–430; E. Dayan, N. Censor, E. R. Buch, M. Sandrini, and L. G. Cohen, "Noninvasive brain stimulation: From physiology to network dynamics and back," *Nature Neuroscience* 16 (2013): 838–844; K. S. Bosley et al., "CRISPR germline engineering—The community speaks," *Nature Biotechnology* 33 (2015): 478–486.

199 **Perhaps nothing:** Over two thousand newspaper stories and combined paper hits between December 25, 2011, and December 25, 2016, were retrieved from a LexisNexis Academic search performed using search terms "hacking" and "brain"—an average of more than four hundred publications per year during this period.

199 **In a 2015 *Atlantic*:** Maria Konnikova, "Hacking the Brain," *Atlantic* (June 2015).

199 **Numerous talks:** Andres Lozano, "Can Hacking the Brain Make You Healthier?" *TED Radio Hour*, National Public Radio, August 9, 2013; Keith Barry, "Brain Magic," TED Conferences, July 21, 2008.

199 **The tone of these:** Greg Gage, Miguel Nicolelis, Tan Le, David Eagleman, Andres Lozano, and Todd Kuiken, "Tech That Can Hack Your Brain," TED

Playlist (6 talks), ted.com/playlists/392/tech_that_can_hack_your_brain (accessed August 1, 2017).

200 **MIT is also famous for:** T. F. Peterson, *Nightwork: A History of Hacks and Pranks at MIT* (Cambridge, MA: MIT Press, 2011).

200 **The most infamous:** G. A. Mashour, E. E. Walker, and R. L. Martuza, "Psychosurgery: Past, present, and future," *Brain Research: Brain Research Reviews* 48 (2005): 409–419.

201 **In one variant:** B. M. Collins and H. J. Stam, "Freeman's transorbital lobotomy as an anomaly: A material culture examination of surgical instruments and operative spaces," *History of Psychology* 18 (2015): 119–131.

201 **Approximately 5 percent:** T. Hilchy. "Dr. James Watts, U.S. Pioneer in Use of Lobotomy, Dies at 90," *New York Times*, November 10, 1994; W. Freeman, "Lobotomy and epilepsy: A study of 1000 patients," *Neurology* 3 (1953): 479–494.

201 **Lobotomies were nevertheless:** G. J. Young et al., "Evita's lobotomy," *Journal of Clinical Neuroscience* 22 (2015): 1883–1888.

202 **Most famous was the case:** Suzanne Corkin, *Permanent Present Tense: The Unforgettable Life of the Amnesic Patient, H. M. (*New York: Basic Books, 2013).

202 **A technique called:** H. Shen, "Neuroscience: Tuning the brain," *Nature* 507 (2014): 290–292; Michael S. Okun and Pamela R. Zeilman, *Parkinson's Disease: Guide to Deep Brain Stimulation Therapy*, National Parkinson Foundation, 2014.

202 **The resulting information:** A. S. Widge et al., "Treating refractory mental illness with closed-loop brain stimulation: Progress towards a patient-specific transdiagnostic approach," *Experimental Neurology* 287 (2017): 461–472.

202 **Brain recordings can:** M. A. Lebedev and M. A. Nicolelis, "Brain-machine interfaces: From basic science to neuroprostheses and neurorehabilitation," *Physiological Reviews* 97 (2017): 767–837.

202 **In an amazing demonstration:** Benedict Carey, "Paralyzed, Moving a Robot with Their Minds," *New York Times*, May 16, 2012.

203 **Controlling a mechanical:** M. K. Manning and A. Irvine, *The DC Comics Encyclopedia* (New York: DK Publishing, 2016).

203 **In one example, researchers:** R. P. Rao et al., "A direct brain-to-brain interface in humans," *PLoS One* 9 (2014): e111332.

203 **Attaching the EEG:** "The Menagerie," directed by Marc Daniels and Robert Butler, *Star Trek*, season 1, episodes 11 and 12, CBS Television, November 17–24, 1966.

203 **In another well-publicized case:** K. N. Kay, T. Naselaris, R. J. Prenger, and J. L. Gallant, "Identifying natural images from human brain activity," *Nature* 452 (2008): 352–355.

203 **"Like computers, human brains":** Tanya Lewis, "How Human Brains Could Be Hacked," LiveScience Blog, livescience.com /37938-how-human-brain -could-be-hacked.html, July 3, 2013.

203 **"Twenty years from now":** Raymond Kurzweil, "Get Ready for Hybrid Thinking," TED Conferences, June 14, 2017.

203 **Kurzweil believes that:** Raymond Kurzweil, *The Singularity Is Near: When Humans Transcend Biology* (New York: Viking, 2005).

204 **Taking a similar tack:** Michio Kaku, *The Future of the Mind: The Scientific Quest to Understand, Enhance, and Empower the Mind* (New York: Anchor Books, 2014).

204 **The Defense Advanced Research:** Biological Technologies Office, "DARPA-BAA-16-33," Defense Advanced Research Projects Agency, 2016.

204 **"We can now see":** Abby Phillip, "A Paralyzed Woman Flew an F-35 Fighter Jet in a Simulator—Using Only Her Mind," *Washington Post*, March 3, 2015.

205 **Even noninvasive brain:** Vanessa Barbara, "Woodpecker to Fix My Brain," *New York Times*, September 27, 2015.

205 **The US presidential election:** John Oliver, "Third Parties," *Last Week Tonight*, HBO, October 16, 2016.

205 **As the founder:** Zoltan Istvan, "Should a Transhumanist Run for US President?" *Huffington Post*, October 8, 2014.

206 **"Who doesn't want":** "Zoltan Istvan and Steve Fuller," Brain Bar Budapest Conference, brainbar.com, June 2, 2016.

206 **The Transhumanist Party:** A. Roussi, "Now This Is an 'Outsider Candidate': Zoltan Istvan, a Transhumanist Running for President, Wants to Make You Immortal," *Salon*, February 19, 2016.

206 **The aspiring politician:** Zoltan Istvan, *The Transhumanist Wager* (Reno, NV: Futurity Imagine Media, 2013).

206 **The transhumanist muse:** Robert Anton Wilson, *Prometheus Rising* (Las Vegas: New Falcon Publications, 1983).

207 **The translation of neuroscience:** D. Martin, "Futurist Known as FM-2030 Is Dead at 69," *New York Times*, July 11, 2000.

207 **In the imaginations:** *The Matrix*, directed by Lana Wachowski and Lilly Wachowski, Warner Bros., 1999; "Q Who," directed by Rob Bowman, *Star Trek: The Next Generation*, season 2, episode 16, CBS Television, May 8, 1989.

207 **Nanobots would be small:** Kevin Shapiro, "This Is Your Brain on Nanobots," *Commentary Magazine*, December 1, 2005.

207 **Although some nanotechnology experts:** A. Moscatelli, "The struggle for control," *Nature Nanotechnology* 8 (2013): 888–890; C. Toumey, "Nanobots today," *Nature Nanotechnology* 8 (2013): 475–476.

207 **Even Nicholas Negroponte:** Nicholas Negroponte, "Nanobots in Your Brain Could Be the Future of Learning," Big Think, bigthink.org/videos/nicholas-negroponte-on-the-future-of-biotech, December 13, 2014.

207 **A digital video series:** *H+: The Digital Series*, directed by Stewart Hendler, youtube.com/user/HplusDigitalSeries, August 8, 2012.

207 **"The upload is":** Natasha Vita-More, quoted in Kevin Holmes, "Talking to the Future Humans: Natasha Vita-More," Vice, October 11, 2011, vice.com/en_us/article/mvpeyq/talking-to-the-future-humans-natasha-vita-more-interview-sex.

208 **To achieve the uploading:** Sebastian Seung, *Connectome: How the Brain's Wiring Makes Us Who We Are* (Boston: Houghton Mifflin Harcourt, 2012).

208 **In Chapter 5:** Price as of June 15, 2017, listed on www.alcor.org.

208 **Alcor is run by:** C. Michallon, "British 'Futurist' Who Runs Cryogenics Facility Says He Plans to Freeze Just His Brain—and Insists His body Is 'Replaceable,'" *Daily Mail*, December 26, 2016.

208 **One of Alcor's early clients:** F. Chamberlain, "A tribute to FM-2030," *Cryonics* 21 (2000): 10–14.

208 **His frozen head:** The Alcor facility in Scottsdale, Arizona, is also where Kim Suozzi's remains are stored.

208 **"Even in cases":** Laura Y. Cabrera, *Rethinking Human Enhancement: Social Enhancement and Emergent Technologies* (New York: Palgrave Macmillan, 2015).

209 **In the arena:** T. Friend, "Silicon Valley's Quest to Live Forever," *New Yorker*, April 3, 2017.

209 **The historian of science:** Thomas S. Kuhn, *The Structure of Scientific Revolutions* (Chicago: University of Chicago Press, 1962).

210 **Among today's humans:** R. Lynn and M. V. Court, "New evidence of dysgenic fertility for intelligence in the United States," *Intelligence* 32 (2004): 193–201.

210 **Beetles, for instance:** O. Béthoux, "The earliest beetle identified," *Journal of Paleontology* 83 (2009): 931–937; N. E. Stork, J. McBroom, C. Gely, and A. J. Hamilton, "New approaches narrow global species estimates for beetles, insects, and terrestrial arthropods," *Proceedings of the National Academy of Sciences* 112 (2015): 7519–7523.

210 **The great biologist:** "Beetlemania," *Economist*, March 18, 2015.

210 **The approach seems:** S. W. Bridge, "The neuropreservation option: Head first into the future," *Cryonics* 16 (1995): 4–7.

210 **"Most of the benefits":** Nick Bostrom, "Superintelligence," BookTV Lecture, C-SPAN, September 12, 2014. Bostrom's estimate of visual data transmission to the brain is considerably larger than the figure I presented in Chapter 6, but it's perhaps based on a quantification of input to the retina rather than retinal output to the brain. The retina contains about a hundred million photoreceptors that detect light from the environment, but this information is compressed dramatically before it leaves the eye. What the brain actually "sees" are the spikes carried by roughly a million ganglion cells per retina, which constitute retinal output only and were the basis for the estimate I cited of ten megabytes per second per eye.

211 **Over four thousand years:** Yoshihide Igarashi, Tom Altman, Mariko Funada, and Barbara Kamiyama, *Computing: A Historical and Technical Perspective* (Boca Raton, FL: CRC, 2014).

211 **The greatest cognitive aid:** Christopher Woods, ed., *Visible Language: Inventions of Writing in the Ancient Middle East and Beyond* (Chicago: University of Chicago Press, 2010).

211 **Philosopher Andy Clark:** Andy Clark, *Supersizing the Mind: Embodiment, Action, and Cognitive Extension* (New York: Oxford University Press, 2011).

211 **The mind and self:** A. Clark and D. J. Chalmers, "The extended mind," *Analysis* 58 (1998): 10–23.

211 **Speaking personally:** In the 2014 US Supreme Court case *Riley v. California*, the court decided that it is unconstitutional to search the contents of a modern cell phone without a warrant. Writing for a unanimous majority, Chief Justice John Roberts described smartphones as "such a pervasive and insistent part of daily life that the proverbial visitor from Mars might conclude they were an important feature of human anatomy."

212 **A different form of cognitive:** Bostonians are proverbially undisciplined drivers.

212 **In this case:** Tim Adams, "Self-Driving Cars: From 2020 You Will Become a Permanent Backseat Driver," *Guardian*, September 13, 2015.

212 **In 1968:** G. S. Brindley and W. S. Lewin, "The sensations produced by electrical stimulation of the visual cortex," *Journal of Physiology* 196 (1968): 479–493.

212 **In the ensuing years:** M. Abrahams, "A Stiff Test for the History Books," *Guardian*, March 16, 2009.

212 **Meanwhile, Brindley's idea:** D. Ghezzi, "Retinal prostheses: Progress toward the next generation implants," *Frontiers in Neuroscience* 9 (2015): 290.

212 **Using a technique:** M. S. Gart, J. M. Souza, and G. A. Dumanian, "Targeted muscle reinnervation in the upper extremity amputee: A technical roadmap," *Journal of Hand Surgery (American Volume)* 40 (2015): 1877–1888.

213 **In 2015, a fifty-nine-year-old man:** E. Cott, "Prosthetic Limbs, Controlled by Thought," *New York Times*, May 20, 2015.

213 **To go beyond rehabilitation:** A. M. Dollar and H. Herr, "Lower extremity exoskeletons and active otheroses: Challenges and state-of-the-art," *IEEE Transactions on Robotics* 24 (2008): 144–158.

213 **In Marvel's *Iron Man* comics:** E. Paul Zehr, "Assembling an Avenger— Inside the Brain of Iron Man," Scientific American Guest Blog, blogs.scientific american.com/guest-blog/assembling-an-avenger-inside-the-brain-of-iron -man, September 26, 2012.

213 **For instance, the HAL-5:** E. Guizzo and H. Goldstein, "The rise of the body bots," *IEEE Spectrum*, October 1, 2005.

214 **In a 2004 essay:** Francis Fukuyama, "Transhumanism," *Foreign Policy* (September/October 2004).

215 **Nootropics also include dietary:** G. Grosso et al., "Omega-3 fatty acids and depression: Scientific evidence and biological mechanisms," *Oxidative Medicine and Cellular Longevity* 2014 (2014): 313570; A. G. Malykh and M. R. Sadaie, "Piracetam and piracetam-like drugs: From basic science to novel clinical applications to CNS disorders," *Drugs* 70 (2010): 287–312.

215 **The most powerful nootropic:** A. Dance, "Smart drugs: A dose of intelligence," *Nature* 531 (2016): S2–3.

215 **Although heavy-hitting:** Margaret Talbot, "Brain Gain," *New Yorker*, April 27, 2009.

215 **A 2005 survey:** S. E. McCabe, J. R. Knight, C. J. Teter, and H. Wechsler, "Non-medical use of prescription stimulants among US college students: Prevalence and correlates from a national survey," *Addiction* 100 (2005): 96–106.

215 **Several studies have questioned:** J. Currie, M. Stabile, and L. E. Jones, "Do stimulant medications improve educational and behavioral outcomes for children with ADHD?" *Journal of Health Economics* 37 (2014): 58–69; I. Ilieva, J. Boland, and M. J. Farah, "Objective and subjective cognitive enhancing effects of mixed amphetamine salts in healthy people," *Neuropharmacology* 64 (2013): 496–505; "AMA Confronts the Rise of Nootropics," American Medical Association, www.ama-assn.org/ama-confronts-rise-nootropics, June 14, 2016.

216 **Silicon Valley start-ups:** T. Amirtha, "Scientists and Silicon Valley Want to Prove Psychoactive Drugs Are Healthy," *Guardian*, February 8, 2016.

216 **Nootrobox, for example:** "Products," Nootrobox, hvmn.com/products (accessed June 15, 2017).

216 **Each ingredient is said:** Nootrobox, Inc., "The Effects of SPRINT, a Combination of Natural Ingredients, on Cognition in Healthy Young Volunteers," Clinical Trials Database, National Institutes of Health, 2016.

216 **"If you aren't taking":** "Alpha Brain," Onnit Labs, June 15, 2017.

216 **Explains businessman and self-help:** Laurie Segall and Erica Fink, "Are Smart Drugs Driving Silicon Valley?" CNN, January 26, 2015.

216 **"All this may be leading":** Talbot, "Brain Gain."

217 **Meanwhile, despite the fact:** Nicholas Kristof, "Overreacting to Terrorism," *New York Times*, March 24, 2016.

218 **They argue that "cognitive-enhancing":** H. Greely et al., "Towards responsible use of cognitive-enhancing drugs by the healthy," *Nature* 456 (2008): 702–705.

218 **Putting the debate:** Expert Group on Cognitive Enhancements, *Boosting Your Brainpower: Ethical Aspects of Cognitive Enhancements*, British Medical Association, 2007.

219 **Particularly given the prevalence:** *Behavioral Health Trends in the United States: Results from the 2014 National Survey on Drug Use and Health*, Substance Abuse and Mental Health Services Administration, 2015.

219 **"Prometheus stole fire":** Ken Goffman (aka R. U. Sirius) and Dan Joy, *Counterculture Through the Ages: From Abraham to Acid House* (New York: Villard Books, 2004).

219 **We can see aspects:** Glenn Greenwald, Ewen MacAskill, and Laura Poitras, "Edward Snowden: The Whistleblower Behind the NSA Surveillance Revelations," *Guardian*, June 11, 2013; Dominic Basulto, "Aaron Swartz and the Rise of the Hacktivist Hero," *Washington Post*, January 14, 2013.

NOTES TO CHAPTER 10

221 **It is the story:** The reader must forgive some poetic license in this chapter.

224 **"Never give in;":** Winston Churchill, "Never Give In, Never (Speech at Harrow)," National Churchill Museum, www.nationalchurchillmuseum.org/never -give-in-never-never-never.html, October 29, 1941.

235 **The ancient Egyptians:** James F. Romano, *Death, Burial, and Afterlife in Ancient Egypt*, Carnegie Series on Egypt (Philadelphia: University of Pennsylvania Press, 1990). Some accounts of ancient Egyptian beliefs describe further components to the soul, including the person's name and representations of the heart and shadow.

235 **The oldest Indian writings:** *The Rig Veda: An Anthology*, translated by Wendy Doniger O'Flaherty (New York: Penguin, 2005).

235 **The Pentateuch gave us:** Matt Stefon, ed., *Judaism: History, Belief, and Practice* (New York: Britannica Educational, 2012); Joshua Dickey, ed., *The Complete Koine-English Reference Bible: New Testament, Septuagint and Strong's Concordance* (Seattle: Amazon Digital Services, 2014) (e-book).

INDEX

CREDIT: LUBA KATZ

ALAN JASANOFF is an award-winning neuroscientist and bioengineer at MIT. He lives in Belmont, Massachusetts.